PRENTICE HALL

ALGEBRA
TOOLS FOR A CHANGING WORLD

Two-Year Algebra Handbook

ISBN: 0-13-432943-0

Printed in the United States of America
1 2 3 4 5 6 7 8   02 01 00 99 98 97

Editorial, design, and production services:
Publishers Resource Group, Inc.

PRENTICE HALL
Simon & Schuster Education Group
A VIACOM COMPANY

Two-Year Algebra Handbook

Contents

Contents (continued)

How to Use the Two-Year Algebra Handbook

The Two-Year Algebra Handbook contains exercises and examples designed to assist you in teaching basic algebraic concepts to students who benefit from learning Algebra over two years. Each lesson in the Student Edition is made up of parts. There is a student worksheet for each part of a lesson.

The Two-Year Algebra Handbook provides you with assistance for
- **Introducing** new concepts
- **Pacing** lessons appropriately
- **Reinforcing** key concepts and vocabulary
- **Reteaching** content to students who need additional instruction to master objectives
- **Assigning** optional exercises

Components of the Two-Year Algebra Handbook

For your convenience and ease of use, the Two-Year Algebra Handbook contains a Pacing and Assignment Guide that corresponds to each chapter and its lessons. The information contained in the guide comes directly from the Student Edition and the Assignment Options in the Teacher's Edition.

Each student worksheet in the handbook includes the following elements.
- **Example**
- **Mini Help**
- **Practice**

The **Example** part of the worksheet may begin with an important fact or rule that will help the student. The example itself directly relates to an example in the part of the lesson to which the worksheet corresponds.

The next part of the worksheet is **Practice**, which includes **Mini Help**. Many of the students in a two-year Algebra program may need to review prerequisite skills. **Mini Help** consists of a few basic exercises in skills related to the concept that is the focus of the worksheet.

The last part of the worksheet contains exercises that allow the student to practice the skills taught in that particular part of the lesson. The exercises are written taking into consideration the needs of students in a two-year Algebra program. The answers to all the exercises are at the back of the handbook.

The Two-Year Algebra Handbook is also an excellent resource for giving practice and review to students in a one-year Algebra program.

The diagram on the next page shows sample pages from the Two-Year Algebra Handbook and the corresponding page in the Student Edition.

Part 1 This icon alerts you to the beginning of **Part 1** of Lesson 3-7. Notice that the subtitle of the Two-Year Algebra Handbook page is the same as the name of the part in the Student Edition.

Part 2 This icon shows where **Part 2** of Lesson 3-7 begins. The subtitle of the next page in the Two-Year Algebra Handbook corresponds to the part name in the Student Edition.

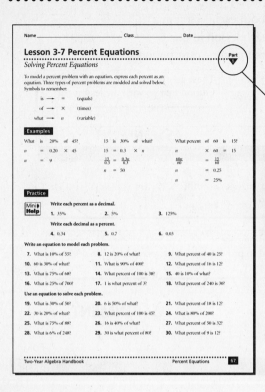

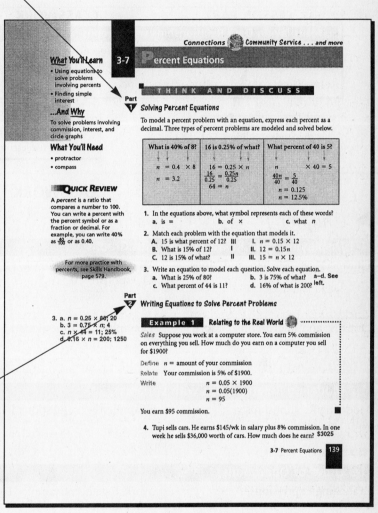

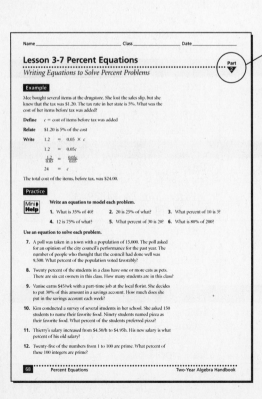

Pacing and Assignment Guide

USING THIS GUIDE

This Pacing and Assignment Guide displays important information about each lesson in a convenient reference form. The guide is organized using the parts of each lesson in the Student Edition. Use the guide to plan how to cover a lesson over more than one class period.

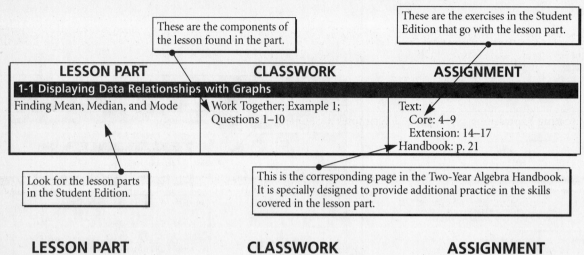

These are the components of the lesson found in the part.

These are the exercises in the Student Edition that go with the lesson part.

Look for the lesson parts in the Student Edition.

This is the corresponding page in the Two-Year Algebra Handbook. It is specially designed to provide additional practice in the skills covered in the lesson part.

LESSON PART	CLASSWORK	ASSIGNMENT

▶ Chapter 1

LESSON PART	CLASSWORK	ASSIGNMENT
1-1 Displaying Data Relationships with Graphs		

Finding Mean, Median, and Mode	Work Together; Example 1; Questions 1–10	Text:
		  Core: 4–9
		  Extension: 14–17
		Handbook: p. 21

Drawing and Interpreting Graphs	Examples 2–3; Questions 11–17	Text:
		  Core: 1–3, 10, 11
		  Extension: 12, 13
		Handbook: p. 22

1-2 Modeling Relationships with Variables		

Modeling Relationships with Variables	Examples 1–2; Questions 1–13	Text:
		  Core: 1–4, 7–18, 20–22
		  Extension: 5, 6, 19, 23–25
		Handbook: p. 23

1-3 Order of Operations		

Evaluating Expressions	Work Together; Examples 1–3; Questions 1–9	Text:
		  Core: 1–5, 7, 9
		  Extension: 13, 14, 29–32
		Handbook: p. 24

Evaluating Expressions with Grouping Symbols	Example 4; Questions 10–13	Text:
		  Core: 6, 8, 10–12, 15–26
		  Extension: 27, 28
		Handbook: p. 25

1-4 Adding and Subtracting Integers		

Adding Integers and Decimals	Work Together; Example 1; Questions 1–12	Text:
		  Core: 3–12, 29–31, 38, 42, 44–45
		  Extension: 1, 2, 32, 33, 46, 51–54
		Handbook: p. 26

Subtracting Integers and Decimals	Examples 2–3; Questions 13–17	Text:
		  Core: 13–28, 36–37, 39–41, 47–50, 55
		  Extension: 34, 35, 43, 56
		Handbook: p. 27

LESSON PART	CLASSWORK	ASSIGNMENT
1-5 Multiplying and Dividing Integers		
Multiplying and Dividing	Work Together; Examples 1–2; Questions 1–13	Text: Core: 1–8, 11, 13–15 Extension: 17, 18, 32–39 Handbook: p. 28
Simplifying Expressions with Exponents	Example 3; Questions 14–15	Text: Core: 4, 9, 10, 12, 16, 19–31 Extension: 40–45, 47, 48 Handbook: p. 29
1-6 Real Numbers and Rational Numbers		
Comparing Rational Numbers	Examples 1–2; Questions 1–7	Text: Core: 1–12, 25–28 Extension: 29, 30, 35 Handbook: p. 30
Evaluating Expressions	Examples 3–5; Questions 8–11	Text: Core: 13–24, 31–34 Extension: 36–45 Handbook: p. 31
1-7 Experimental Probability and Simulations		
Finding Experimental Probability	Work Together; Example 1; Questions 1–5	Text: Core: 1–5, 7–10 Extension: 6, 11 Handbook: p. 32
Conducting a Simulation	Example 2; Questions 6–7	Text: Core: 12–19 Extension: 20–22 Handbook: p. 33
1-8 Organizing Data into Matrices		
Organizing Data into Matrices	Examples 1–2; Questions 1–7	Text: Core: 1–18, 24–26 Extension: 19, 20–23 Handbook: p. 34
1-9 Variables and Formulas in Spreadsheets		
Variables and Formulas in Spreadsheets	Examples 1–2; Questions 1–6	Text: Core: 3–5, 7–15 Extension: 1, 2, 6, 16–19 Handbook: p. 35

▶ *Chapter 2*

LESSON PART	CLASSWORK	ASSIGNMENT
2-1 Analyzing Data Using Scatter Plots		
Drawing and Interpreting Scatter Plots	Work Together; Example 1; Questions 1–7	Text: Core: 17, 19, 20, 22 Extension: 11, 18, 21 Handbook: p. 36
Analyzing Trends in Data	Example 2; Questions 8–9	Text: Core: 1–8, 12–15 Extension: 9, 10, 16 Handbook: p. 37
2-2 Relating Graphs to Events		
Interpreting Graphs	Example 1; Questions 1–3	Text: Core: 17–19 Extension: 5 Handbook: p. 38
Sketching Graphs	Example 2; Questions 4–5	Text: Core: 1-4 Extension: 16 Handbook: p. 39

LESSON PART	CLASSWORK	ASSIGNMENT

Classifying Data	Example 3; Questions 1–6; Work Together	Text:
		  Core: 7–14
		  Extension: 15
		Handbook: p. 40

2-3 Linking Graphs to Tables		

Linking Graphs to Tables	Work Together; Example; Questions 1–10	Text:
		  Core: 1-9, 11, 12
		  Extension: 10, 13, 14
		Handbook: p. 41

2-4 Functions		

Identifying Relations and Functions	Example 1; Questions 1–3	Text:
		  Core: 9–12, 18–21
		  Extension: 38, 51–53
		Handbook: p. 42

Evaluating Functions	Examples 2–3; Questions 4–6	Text:
		  Core: 1–8, 25–37, 39–46
		  Extension: 47–50
		Handbook: p. 43

Analyzing Graphs	Questions 7–9; Work Together	Text:
		  Core: 13–17, 24
		  Extension: 22, 23
		Handbook: p. 44

2-5 Writing a Function Rule		

Understanding Function Notation	Questions 1–2	Text:
		  Core: 1–8, 18–25
		  Extension: 10, 30
		Handbook: p. 45

Using a Table of Values	Example 1; Questions 3–4	Text:
		  Core: 12–14, 26–29, 32–34
		  Extension: 11, 31
		Handbook: p. 46

Using Words to Write a Rule	Examples 2–3; Questions 5–8	Text:
		  Core: 15–17
		  Extension: 9
		Handbook: p. 47

2-6 The Three Views of a Function		

The Three Views of a Function	Examples 1–3; Questions 1–8	Text:
		  Core: 1–8, 12–30
		  Extension: 9–11, 31–35
		Handbook: p. 48

2-7 Families of Functions		

Identifying the Family of an Equation	Work Together; Example 1; Questions 1–7	Text:
		  Core: 1–12, 19–22
		  Extension: 13–18, 38
		Handbook: p. 49

Identifying the Family of a Graph	Example 2; Questions 8–9	Text:
		  Core: 23–33
		  Extension: 34–37, 39, 40
		Handbook: p. 50

2-8 The Probability Formula		

Finding Theoretical Probability	Examples 1–2; Questions 1–5	Text:
		  Core: 1–9, 12, 15–21, 23–30, 41, 42
		  Extension: 10, 11, 22, 43
		Handbook: p. 51

Using a Tree Diagram to Find a Sample Space	Example 3; Question 6	Text:
		  Core: 13, 31–38
		  Extension: 14, 39, 40
		Handbook: p. 52

LESSON PART	CLASSWORK	ASSIGNMENT

▶ *Chapter 3*

3-1 Modeling and Solving Equations

LESSON PART	CLASSWORK	ASSIGNMENT

Solving Addition and Subtraction Equations	Example 1; Questions 1–3	Text:
		  Core: 4, 5, 7, 12
		  Extension: 10, 11, 28, 29
		Handbook: p. 53

Solving Multiplication and Division Equations	Example 2; Questions 4–8	Text:
		  Core: 3, 6, 16–27, 39–50
		  Extension: 8, 9, 30, 31
		Handbook: p. 54

Modeling by Writing Equations	Example 3; Question 9	Text:
		  Core: 1, 2, 14, 15, 32–38, 51, 52, 54
		  Extension: 13, 53, 72
		Handbook: p. 55

3-2 Modeling and Solving Two-Step Equations

LESSON PART	CLASSWORK	ASSIGNMENT

Using Tiles	Work Together; Example 1; Questions 1–4	Text:
		  Core: 1–11
		  Extension: 24–26, 50, 51, 60, 61
		Handbook: p. 56

Using Properties	Examples 2–3; Questions 5–9	Text:
		  Core: 12–23, 27≠46, 52–59
		  Extension: 47–49
		Handbook: p. 57

3-3 Combining Like Terms to Solve Equations

LESSON PART	CLASSWORK	ASSIGNMENT

Combing Like Terms	Work Together; Example 1; Questions 1–4	Text:
		  Core: 1, 3–13
		  Extension: 2, 24, 25
		Handbook: p. 58

Solving Equations	Examples 2–3; Questions 5–6	Text:
		  Core: 14–21, 26–39
		  Extension: 22, 23, 40–49
		Handbook: p. 59

3-4 Using the Distributive Property

LESSON PART	CLASSWORK	ASSIGNMENT

Simplifying Variable Expressions	Work Together; Questions 1–4	Text:
		  Core: 1–3, 6–7, 43–48
		  Extension: 4, 5, 18, 19, 20, 42
		Handbook: p. 60

Solving and Modeling Equations	Examples 1–2; Questions 5–6	Text:
		  Core: 21–36
		  Extension: 37–41
		Handbook: p. 61

3-5 Rational Numbers and Equations

LESSON PART	CLASSWORK	ASSIGNMENT

Multiplying by a Reciprocal	Example 1; Questions 1–3	Text:
		  Core: 1–8, 14, 15, 17, 19–27, 54
		  Extension: 10–13, 55
		Handbook: p. 62

Multiplying by a Common Denominator	Examples 2–3; Questions 4–6	Text:
		  Core: 16, 18, 28–53
		  Extension: 9, 37, 38, 56, 57
		Handbook: p. 63

3-6 Using Probability

LESSON PART	CLASSWORK	ASSIGNMENT

Finding the Probability of Independent Events	Work Together; Example 1; Questions 1– 5	Text:
		  Core: 1–3, 5, 7–16, 30, 37
		  Extension: 4
		Handbook: p. 64

© Prentice-Hall, Inc.

LESSON PART	CLASSWORK	ASSIGNMENT
Finding the Probability of Dependent Events	Example 2; Questions 6–7	Text: Core: 6, 17–22 Extension: 23, 38 Handbook: p. 65
Finding Probability Using and Equations	Example 3; Question 8	Text: Core: 26, 27, 29 Extension: 24, 25, 28, 39 Handbook: p. 66
3-7 Percent Equations		
Solving Percent Equations	Questions 1–3	Text: Core: 1–9, 15–26, 31–35 Extension: 14, 30 Handbook: p. 67
Writing Equations to Solve Percent Problems	Examples 1–3, Questions 4–7	Text: Core: 13, 27, 28 Extension: 10, 12 Handbook: p. 68
Simple Interest	Example 4, Questions 8–9; Work Together	Text: Core: 36–39 Extension: 11, 29 Handbook: p. 69
3-8 Percent of Change		
Percent of Change	Examples 1–2; Questions 1–4	Text: Core: 1–20, 22–34, 39–44 Extension: 21, 35–38, 45 Handbook: p. 70

▶ *Chapter 4*

LESSON PART	CLASSWORK	ASSIGNMENT
4-1 Using Proportions		
Using Properties of Equality	Work Together; Example 1; Questions 1–3	Text: Core: 24–31 Extension: 38, 39 Handbook: p. 71
Using Cross Products	Examples 2–3; Questions 4–6	Text: Core: 7–8, 32–35 Extension: 23, 40–42 Handbook: p. 72
Solving Percent Problems Using Proportions	Example 4; Questions 7–8	Text: Core: 9–20 Extension: 21, 22, 36, 37, 43 Handbook: p. 73
4-2 Equations with Variables on Both Sides		
Using Tiles to Solve Equations	Work Together; Example 1; Questions 1–8	Text: Core: 1, 3–6, 22–24, 44–46 Extension: 2, 7, 32 Handbook: p. 74
Using Properties of Equality	Examples 2–3; Questions 9–11	Text: Core: 14, 15, 25–27, 47–49 Extension: 34, 35, 42, 43, 53 Handbook: p. 75
Solving Special Types of Equations	Examples 4–5; Questions 12–14	Text: Core: 8–13, 16–21, 28–30, 50–52 Extension: 31, 33, 36–41 Handbook: p. 76

© Prentice-Hall, Inc.

LESSON PART	CLASSWORK	ASSIGNMENT
4-3 Solving Absolute Value Equations		
Solving Absolute Value Equations	Work Together; Examples 1–3; Questions 1–10	Text: Core: 1–8, 11–25 Extension: 9, 33–40 Handbook: p. 77
Modeling by Writing Equations	Example 4; Question 11	Text: Core: 26, 27, 29–32, 41, 42 Extension: 10, 28, 43, 44 Handbook: p. 78
4-4 Transforming Formulas		
Transforming Formulas	Examples 1–4; Questions 1–6	Text: Core: 1–8, 10–23 Extension: 9, 24–27 Handbook: p. 79
4-5 Solving Inequalities Using Addition and Subtraction		
Graphing and Writing Inequalities	Work Together; Questions 1–7	Text: Core: 1–6, 8–12 Extension: 7, 16–21 Handbook: p. 80
Using Addition to Solve Inequalities	Example 1; Questions 8–10	Text: Core: 13, 15, 25–36 Extension: 50, 51–56 Handbook: p. 81
Using Subtraction to Solve Inequalities	Example 2; Questions 11–14	Text: Core: 14, 23, 24, 37–48 Extension: 22, 49, 57 Handbook: p. 82
4-6 Solving Inequalities Using Multiplication and Division		
Solving Inequalities Using Multiplication	Work Together; Examples 1–2; Questions 1–7	Text: Core: 5–10, 17–22, 32–39 Extension: 25, 30, 44–46 Handbook: p. 83
Solving Inequalities Using Division	Example 3; Questions 8–11	Text: Core: 1–4, 11–16, 31, 40–43 Extension: 23, 24, 26–29 Handbook: p. 84
4-7 Solving Multi-Step Inequalities		
Solving with Variables on One Side	Examples 1–2; Questions 1–2	Text: Core: 1–7, 9–15, 17–18 Extension: 16, 28, 38–40 Handbook: p. 85
Solving with Variables on Both Sides	Example 3; Questions 3–5	Text: Core: 8, 20–26, 29–37 Extension: 27, 41–43 Handbook: p. 86
4-8 Compound Inequalities		
Solving Compound Inequalities Joined by *And*	Examples 1–2; Questions 1–4	Text: Core: 1, 3, 5, 7, 13–14, 21–23 Extension: 25–26, 45–48 Handbook: p. 87
Solving Compound Inequalities Joined by *Or*	Example 3; Questions 5–6	Text: Core: 2, 4, 6, 8, 15–20, 24 Extension: 27–28, 49–51 Handbook: p. 88

© Prentice-Hall, Inc.

LESSON PART	CLASSWORK	ASSIGNMENT

Solving Absolute Value Inequalities	Example 4; Questions 7–11	Text:
		  Core: 29–44
		  Extension: 9–12
		Handbook: p. 89

4-9 Interpreting Solutions

LESSON PART	CLASSWORK	ASSIGNMENT

Solving Inequalities Given a Replacement Set	Example 1; Questions 1–2	Text:
		  Core: 1–16
		  Extension: 24–29
		Handbook: p. 90

Determining a Reasonable Answer	Example 2; Questions 3–4	Text:
		  Core: 17–20
		  Extension: 21–23
		Handbook: p. 91

▶ Chapter 5

5-1 Slope

LESSON PART	CLASSWORK	ASSIGNMENT

Counting Units to Find Slope	Work Together; Examples 1–2; Questions 1–4	Text:
		  Core: 1, 21–23
		  Extension: 2, 24, 31
		Handbook: p. 92

Using Coordinates to Find Slope	Examples 3–4; Questions 5–7	Text:
		  Core: 3–8, 25–27
		  Extension: 28–30, 32, 33
		Handbook: p. 93

Graphing a Line Given Its Slope and a Point	Example 5; Questions 8–9	Text:
		  Core: 9–16
		  Extension: 17–20
		Handbook: p. 94

5-2 Rates of Change

LESSON PART	CLASSWORK	ASSIGNMENT

Finding Rate of Change	Example 1; Questions 1–3	Text:
		  Core: 1–6
		  Extension: 22, 24
		Handbook: p. 95

Using a Table	Example 2; Question 4	Text:
		  Core: 7, 12–17
		  Extension: 18
		Handbook: p. 96

Linear Functions	Example 3; Questions 5–7	Text:
		  Core: 8–11, 19, 20
		  Extension: 21, 23
		Handbook: p. 97

5-3 Direct Variation

LESSON PART	CLASSWORK	ASSIGNMENT

Direct Variation	Example 1; Questions 1–4	Text:
		  Core: 2–7, 25–32
		  Extension: 1, 8, 33
		Handbook: p. 98

Using the Constant of Variation to Write Equations	Examples 2–3; Questions 5–6	Text:
		  Core: 9–12, 22
		  Extension: 19–21, 23
		Handbook: p. 99

Using Proportions	Example 4; Questions 7–8	Text:
		  Core: 13–17
		  Extension: 18, 24
		Handbook: p. 100

© Prentice-Hall, Inc.

LESSON PART	CLASSWORK	ASSIGNMENT
5-4 Slope-Intercept Form		
Defining Slope-Intercept Form	Work Together; Examples 1–2; Questions 1–15	Text: Core: 1–6, 12–20 Extension: 10, 11 Handbook: p. 101

Writing Equations	Example 3	Text:
		  Core: 7–9, 21–23, 27–30
		  Extension: 24–26
		Handbook: p. 102

5-5 Writing the Equation of a Line		

Writing the Equation of a Line	Examples 1–3; Questions 1–4	Text:
		  Core: 1–12, 14–22, 24–28
		  Extension: 13, 23, 29, 30
		Handbook: p. 103

5-6 Scatter Plots and Equations of Lines		

Trend Line	Example 1; Question 1	Text:
		  Core: 1–6, 8–10, 19
		  Extension: 7, 18
		Handbook: p. 104

Line of Best Fit	Example 2;
	Questions 2–9; Work Together
		Text:
		  Core: 11–14, 16
		  Extension: 15, 17
		Handbook: p. 105

5-7 $Ax + By = C$ Form		

Graphing Equations	Examples 1–2; Questions 1–3	Text:
		  Core: 1–11, 15–22
		  Extension: 12–14, 23
		Handbook: p. 106

Writing Equations	Examples 3–4; Questions 4–5	Text:
		  Core: 24–31
		  Extension: 32
		Handbook: p. 107

5-8 Parallel and Perpendicular Lines		

Parallel Lines	Work Together; Example 1;
	Questions 1–9
		Text:
		  Core: 1–12, 25–30
		  Extension: 32, 33, 37–40
		Handbook: p. 108

Perpendicular Lines	Example 2; Questions 10–14	Text:
		  Core: 13–24, 41–50
		  Extension: 31, 34–36, 51–53
		Handbook: p. 109

5-9 Using the x-intercept		

Using the x-intercept	Examples 1–3; Questions 1–5	Text:
		  Core: 1–17, 22–25
		  Extension: 18–21, 26–29
		Handbook: p. 110

▶ *Chapter 6*

LESSON PART	CLASSWORK	ASSIGNMENT
6-1 Solving Systems by Graphing		

Solving Systems with One Solution	Examples 1–2; Questions 1–4	Text:
		  Core: 1–5, 7–10
		  Extension: 6, 11, 12
		Handbook: p. 111

Solving Special Types of Systems	Example 3; Questions 5–10;
	Work Together
		Text:
		  Core: 13–24, 29–36, 41
		  Extension: 25–28, 37–40, 42–46
		Handbook: p. 112

LESSON PART	CLASSWORK	ASSIGNMENT
6-2 Solving Systems Using Substitution		

Solving Systems with One Solution	Work Together; Examples 1–2; Questions 1–9	Text:
		   Core: 9, 11–22, 29–33, 39
		   Extension: 10, 28, 38
		Handbook: p. 113

Solving Special Types of Systems	Example 3; Questions 10–14	Text:
		   Core: 1–8, 23–27, 34–37
		   Extension: 40–48
		Handbook: p. 114

6-3 Solving Systems Using Elimination		

Adding or Subtracting Equations	Examples 1–2; Questions 1–3	Text:
		   Core: 1–5, 10–17, 30–33
		   Extension: 26, 27, 38
		Handbook: p. 115

Multiplying First	Example 3; Questions 4–9;
	Work Together
		Text:
		   Core: 6–8, 18–25, 34–37
		   Extension: 9, 28, 29, 39
		Handbook: p. 116

6-4 Writing Systems		

Writing Systems	Examples 1–2; Questions 1–4	Text:
		   Core: 5–15, 19–24
		   Extension: 1–4, 16–18, 25, 26
		Handbook: p. 117

6-5 Linear Inequalities		

Linear Inequalities	Work Together; Examples 1–3;
	Questions 1-12
		Text:
		   Core; 1–6, 8–24. 27–30, 36–39
		   Extension: 7, 25, 26, 31–35, 40
		Handbook: p. 118

6-6 Systems of Linear Inequalities		

Systems of Linear Inequalities	Work Together; Examples 1–2;
	Questions 1–15
		Text:
		   Core: 1–3, 5–16, 23–27
		   Extension: 4, 17,–22
		Handbook: p. 119

6-7 Concepts of Linear Programming		

Concepts of Linear Programming	Work Together; Examples 1–3;
	Questions 1–7
		Text:
		   Core: 1–11
		   Extension: 12–15
		Handbook: p. 120

6-8 Systems of Nonlinear Equations		

Systems of Nonlinear Equations	Examples 1–3; Questions 1–4	Text:
		   Core: 1–8, 10–13, 15, 20–31
		   Extension: 9, 14, 16–19
		Handbook: p. 121

▶ *Chapter 7*

LESSON PART	CLASSWORK	ASSIGNMENT
7-1 Exploring Quadratic Functions		

Quadratic Functions	Work Together; Questions 1–5	Text:
		   Core: 1–4, 17–20
		   Extension: 21–24, 29–33
		Handbook: p. 122

The Role of a	Examples 1–2; Questions 6–11	Text:
		   Core: 5–16, 25–28
		   Extension: 34–39
		Handbook: p. 123

© Prentice-Hall, Inc.

LESSON PART	CLASSWORK	ASSIGNMENT
7-2 Graphing Simple Quadratic Functions		
Graphing Simple Quadratic Functions	Work Together; Examples 1–2; Questions 1–8	Text: Core: 2–17, 21–33 Extension: 1, 18–20, 34 Handbook: p. 124
7-3 Graphing Quadratic Functions		
Graphing $y = ax^2 + bx + c$	Work Together; Examples 1–2; Question 1–3	Text: Core: 1–27 Extension: 28–30 Handbook: p. 125
Quadratic Inequalities	Example 3; Questions 4–6	Text: Core: 31–39 Extension: 40–42 Handbook: p. 126
7-4 Square Roots		
Finding Square Roots	Example 1; Questions 1–2	Text: Core: 1–16, 37, 52–59 Extension: 35, 36, 50, 51 Handbook: p. 127
Estimating and Using Square Roots	Examples 2–4; Questions 3–5	Text: Core: 17–20, 23–34, 38–49 Extension: 21, 22, 60–69 Handbook: p. 128
7-5 solving Quadratic Equations		
Using Square Roots to Solve Equations	Work Together; Examples 1–2; Questions 1–13	Text: Core: 13–19, 37–42 Extension: 20, 21, 32–35 Handbook: p. 129
Finding the Number of Solutions	Example 3; Questions 14–15	Text: Core: 1–12, 22–29 Extension: 30, 31, 36 Handbook: p. 130
7-6 Using the Quadratic Formula		
Using the Quadratic Formula	Examples 1–3; Questions 1–7	Text: Core: 1–18, 21–32 Extension: 19, 20, 33–35 Handbook: p. 131
7-7 Using the Discriminant		
Using the Discriminant	Work Together; Examples 1–2; Questions 1–11	Text: Core: 1–15, 22–30, 32–39 Extension: 16–21, 31, 40, 41 Handbook: p. 132

▶ *Chapter 8*

LESSON PART	CLASSWORK	ASSIGNMENT
8-1 Exploring Exponential Functions		
Exploring Exponential Patterns	Work Together; Example 1; Questions 1–6	Text: Core: 1, 2, 23 Extension: 17, 18, 23–25 Handbook: p. 133
Evaluating Exponential Functions	Example 2; Questions 7–8	Text: Core: 3–14, 19–21 Extension: 15, 22 Handbook: p. 134
Graphing Exponential Functions	Example 3; Question 9	Text: Core: 26–31 Extension: 16 Handbook: p. 135

© Prentice-Hall, Inc.

LESSON PART	CLASSWORK	ASSIGNMENT
8-2 Exponential Growth		

Modeling Exponential Growth	Work Together; Example 1; Questions 1–10	Text:
		   Core: 1–14, 23–29
		   Extension: 15, 16, 19, 22, 30
		Handbook: p. 136

Finding Compound Interest	Examples 2–3; Questions 11–13	Text:
		   Core: 17, 21, 31–34
		   Extension: 18, 20
		Handbook: p. 137

8-3 Exponential Decay		

Exponential Decay	Examples 1–3; Questions 1–8	Text:
		   Core: 1–15, 26–28, 31–36
		   Extension: 16, 25, 29, 30, 37–39
		Handbook: p. 138

8-4 Zero and Negative Exponents		

Using Zero and Negative Integers as Exponents	Work Together; Examples 1–3; Questions 1–10	Text:
		   Core: 1–18, 20–37, 52–65
		   Extension: 19, 38–43
		Handbook: p. 139

Relating the Properties to Exponential Functions	Example 4; Questions 11–12	Text:
		   Core: 44–49, 73–75
		   Extension: 50, 51, 66–72
		Handbook: p. 140

8-5 Scientific Notation		

Writing Numbers in Scientific Notation	Example 1; Questions 1–6	Text:
		   Core: 1–4, 10–30
		   Extension: 31, 47–53
		Handbook: p. 141

Calculating with Scientific Notation	Examples 2–3; Questions 7–12	Text:
		   Core: 5–9, 32–36, 37–42
		   Extension: 43–46
		Handbook: p. 142

8-6 A Multiplication Property of Exponents		

Multiplying Powers	Work Together; Example 1; Questions 1–9	Text:
		   Core: 1–11, 20–47
		   Extension: 19, 48–50
		Handbook: p. 143

Working with Scientific Notation	Examples 2–3; Questions 10–12	Text:
		   Core: 12–17, 55–57
		   Extension: 18, 51–54, 58, 59
		Handbook: p. 144

8-7 More Multiplication Properties of Exponents		

Raising a Power to a Power	Example 1; Questions 1–6	Text:
		   Core: 1–28, 41–46
		   Extension: 29, 47–50
		Handbook: p. 145

Raising a Product to a Power	Examples 2–3; Questions 7–10	Text:
		   Core: 30–36, 52–55
		   Extension: 37–40, 51
		Handbook: p. 146

8-8 Division Properties of Exponents		

Dividing Powers with the Same Base	Example 1–2; Questions 1–5	Text:
		   Core: 1–12, 19–2
		   Extension: 13–18, 24–26
		Handbook: p. 147

© Prentice-Hall, Inc.

LESSON PART	CLASSWORK	ASSIGNMENT
Raising a Quotient to a Power	Example 3; Questions 6–11	Text: Core: 27–50 Extension: 51–58 Handbook: p. 148

▶ *Chapter 9*

9-1 The Pythagorean Theorem

Solving Equations Using the Pythagorean Theorem	Work Together; Examples 1–2; Questions 1–9	Text: Core: 1–12, 14–17 Extension: 13, 18–20 Handbook: p. 149
Using the Converse	Example 3; Question 10	Text: Core: 21–36 Extension: 37–39 Handbook: p. 150

9-2 The Distance Formula

Find the Distance	Work Together; Examples 1–2; Questions 1–7	Text: Core: 1–15 Extension: 16, 17, 31–34 Handbook: p. 151
Using the Midpoint Formula	Example 3; Questions 8–9	Text: Core: 18–29 Extension: 30, 35, 36 Handbook: p. 152

9-3 Trigonometric Ratios

Finding Trigonometric Ratios	Work Together; Examples 1–2; Questions 1–5	Text: Core: 1–18 Extension: 23–25 Handbook: p. 153
Solving Problems Using Trigonometric Ratios	Example 3; Question 6	Text: Core: 19–22 Extension: 26–28 Handbook: p. 154

9-4 Simplifying Radicals

Multiplication with Radicals	Examples 1–3; Questions 1–3	Text: Core: 1–20, 40–47 Extension: 21–24, 56 Handbook: p. 155
Division with Radicals	Examples 4–6; Questions 4–6	Text: Core: 25–34, 48–55 Extension: 35–39 Handbook: p. 156

9-5 Adding and Subtracting Radicals

Simplifying Sums and Differences	Work Together; Examples 1–2; Questions 1–6	Text: Core: 1–15, 17–25 Extension: 16, 39–41 Handbook: p. 157
Simplifying Products, Sums, and Differences	Examples 3–4; Questions 7–9	Text: Core: 26–34 Extension: 35–38, 42, 43 Handbook: p. 158

LESSON PART	CLASSWORK	ASSIGNMENT
9-6 Solving Radical Equations		

Solving a Radical Equation	Examples 1–3; Questions 1–10	Text:
		  Core: 1–6, 9–17
		  Extension: 7, 8, 18–20
		Handbook: p. 159

Solving Equations with Extraneous Solutions	Example 4; Question 11	Text:
		  Core: 21–29, 32–37
		  Extension: 30, 31
		Handbook: p. 160

9-7 Graphing Square Root Functions		

Graphing Square Root Functions	Examples 1–3; Questions 1–13	Text:
		  Core: 1–9, 11–13, 15–26
		  Extension: 10, 14, 27–31
		Handbook: p. 161

9-8 Analyzing Data Using Standard Deviation		

Analyzing Data Using Standard Deviation	Work Together; Examples 1–2; Questions 1–10	Text:
		  Core: 1–6, 11–14
		  Extension: 7–10, 15, 16
		Handbook: p. 162

▶ *Chapter 10*

LESSON PART	CLASSWORK	ASSIGNMENT
10-1 Adding and Subtracting Polynomials		

Describing Polynomials	Work Together; Example 1, Questions 1–5	Text:
		  Core: 1–13
		  Extension: 23, 24, 29, 30
		Handbook: p. 163

Adding Polynomials	Example 2; Questions 6–7	Text:
		  Core: 14, 15, 25–28
		  Extension: 31–33
		Handbook: p. 164

Subtracting Polynomials	Example 3; Questions 8–9	Text:
		  Core: 16–21, 34–39
		  Extension: 22, 40
		Handbook: p. 165

10-2 Multiplying and Factoring		

Multiplying by a Monomial	Example 1; Questions 1–3	Text:
		  Core: 1–5, 9–20
		  Extension: 33, 51
		Handbook: p. 166

Factoring Out a Monomial	Example 2–5; Questions 4–8	Text:
		  Core: 6–8, 21–32, 36–47
		  Extension: 34, 35, 48–50
		Handbook: p. 167

10-3 Multiplying Polynomials		

Multiplying Two Binomials	Example 1; Questions 1–4	Text:
		  Core: 1–6
		  Extension: 7–10
		Handbook: p. 168

Multiplying Using FOIL	Examples 2–3; Questions 5–8	Text:
		  Core: 23–30
		  Extension: 20–22, 31, 32
		Handbook: p. 169

Multiplying a Trinomial and a Binomial	Example 4; Question 9	Text:
		  Core: 11–19
		  Extension: 33–36
		Handbook: p. 170

© Prentice-Hall, Inc.

LESSON PART	CLASSWORK	ASSIGNMENT
10-4 Factoring Trinomials		
Using Tiles	Work Together; Example 1; Questions 1–5	Text: Core: 1–15 Extension: 22, 23 Handbook: p. 171
Testing Possible Factors	Examples 2–3; Questions 6–7	Text: Core: 24–39 Extension: 16–21 Handbook: p. 172
Factoring $ax^2 + bx + c$	Example 4; Question 8	Text: Core: 43–54 Extension: 40–42 Handbook: p. 173
10-5 Factoring Special Cases		
Factoring a Difference of Two Squares	Work Together; Examples 1–2; Questions 1–7	Text: Core: 25–32 Extension: 1–4, 37, 38 Handbook: p. 174
Factoring a Perfect Square Trinomial	Examples 3–5; Questions 8–12	Text: Core: 5–20 Extension: 21–24, 33–36 Handbook: p. 175
10-6 Solving Equations by Factoring		
Solving Equations by Factoring	Examples 1–3; Questions 1–4	Text: Core: 1–15, 18–29, 35–40 Extension: 16, 17, 30–34, 41–45 Handbook: p. 176
10-7 Choosing an Appropriate Method for Solving		
Choosing an Appropriate Method for Solving	Examples 1–4; Questions 1–8	Text: Core: 1–10, 13–28 Extension: 11, 12, 29–33 Handbook: p. 177

▶ *Chapter 11*

LESSON PART	CLASSWORK	ASSIGNMENT
11-1 Inverse Variation		
Solving Inverse Variations	Work Together; Example 1; Questions 1–8	Text: Core: 3–14, 16–22 Extension: 1, 2, 15, 23–25 Handbook: p. 178
Comparing Direct and Indirect Variations	Example 2, Question 9	Text: Core: 26–34 Extension: 35, 36 Handbook: p. 179
11-2 Rational Functions		
Exploring Rational Functions	Questions 1–3	Text: Core: 1–8 Extension: 9–12, 26 Handbook: p. 180
Graphing Rational Functions	Examples 1–2; Questions 4–12; Work Together	Text: Core: 13–24, 29–44 Extension: 25, 27–28 Handbook: p. 181

LESSON PART	CLASSWORK	ASSIGNMENT
11-3 Rational Expressions		

Simplifying Rational Expressions	Work Together; Examples 1–2; Questions 1–8	Text:
		  Core: 1–16
		  Extension: 17–19
		Handbook: p. 182

Multiplying and Dividing Rational Expressions	Examples 3–4; Questions 9–13	Text:
		  Core: 20–36
		  Extension: 37–39
		Handbook: p. 183

11-4 Operations with Rational Expressions		

Operations with Rational Expressions	Work Together; Examples 1–4; Questions 1–8	Text:
		  Core: 1–36
		  Extension: 37–51
		Handbook: p. 184

11-5 Solving Rational Equations		

Solving Rational Equations	Examples 1–3, Questions 1–3	Text:
		  Core: 2–15, 20–29
		  Extension: 1, 16–19, 30–31
		Handbook: p. 185

11-6 Counting Outcomes and Permutations		

Using the Multiplication Counting Principle	Work Together; Example 1; Questions 1–5	Text:
		  Core: 1–5, 26, 27
		  Extension: 18, 28
		Handbook: p. 186

Finding Permutations	Examples 2–3; Questions 6–9	Text:
		  Core: 6–17, 22–25
		  Extension: 19–21
		Handbook: p. 187

11-7 Combinations		

Combinations	Examples 1–2, Questions 1–4	Text:
		  Core: 1–15, 20–28, 30–33
		  Extension: 16–19, 29, 34
		Handbook: p. 188

Lesson 1-1
Displaying Data Relationships with Graphs

Finding Mean, Median, and Mode

Example

Find the mean, median, and mode.

Age of Employees at a Local Fast Food Restaurant

```
                x
        x   x   x       x
x   x   x   x       x           x       x
16  17  18  19  20  21  22  23  24  25  26
                    Age
```

Mean: $\dfrac{16 + 2(17) + 3(18) + 2(19) + 2(21) + 24 + 26}{12}$ = 19.5

Median: 16   17   17   18   18   18   19   19   21   21   24   26

$\dfrac{18 + 19}{2}$ = 18.5

Mode: 18 (the data item that appears most often)

Practice

Order each group from least to greatest.

1. 1 3 6 2 4 5 7 9   **2.** 3 5 3 4 2 0 −2 1   **3.** −23 −15 −6 −14 −4

4. 17 8.5 65 47.5 20   **5.** 1.7 0.9 1.3 1.5 1.0   **6.** 75 50.5 31 2.3 −13

7. Twelve people at the library were asked how many books they planned to check out. Their responses are displayed below. Find the mean, median, and mode.

Number of Library Books to Be Checked Out

```
            x
        x   x   x   x
x       x   x   x   x   x       x
0   1   2   3   4   5   6   7   8
```

8. The heights, in inches, of members of a girls' basketball team are listed below. Draw a line plot of the data.
68 66 63 64 68 69 64 71 61 68 66 69

9. Find the mean, median, and mode for the data in Exercise 8.

10. Suppose a player who is 68 in. tall quits the team. Her replacement is 66 in. tall. How does this change affect the mode, median, and mean?

Lesson 1-1
Displaying Data Relationships with Graphs

Part **2**

Drawing and Interpreting Graphs

Example

Draw a multiple bar graph for the travel cost data.

Travel Costs in the U.S.

	Food	Hotel	Car Rental
West	$59.25	$90.74	$46.55
Midwest	$54.43	$80.68	$54.21
South	$53.85	$79.16	$49.21
North	$53.85	$102.91	$54.96
U.S. Average	$56.39	$85.88	$50.68

Source: *USA Today*

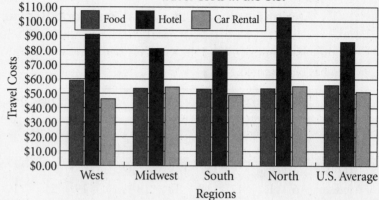

The highest cost is $102.91. So a reasonable range for the vertical scale is from 0 to $110 with every $10 labeled on the axis.

To draw a bar on the graph, estimate its placement based based on the vertical scale.

A multiple bar graph must include a key.

Practice

Mini ▶ Help

Answer the following questions using the graph above.

1. Where are hotel costs lowest?

2. Where is car rental cost highest?

3. Where are food costs higher than the U.S. average?

4. Where are hotel costs lower than the U.S. average?

5. a. Use a triple bar graph or a triple line graph to display the data at right.

   b. Why did you choose the type of graph you did?

   c. What trends do you see in the number of cars produced?

   d. The United States had an economic recession in the late 1980s and early 1990s. How is the recession reflected in the data?

U.S. Passenger Car Production (millions)

Year	Chrysler	Ford	General Motors
1970	1.3	2.0	3.0
1975	0.9	1.8	3.7
1980	0.6	1.3	4.1
1985	1.3	1.6	4.9
1990	0.7	1.4	2.8

Source: *Information Please Almanac 1993*

Lesson 1-2 Modeling Relationships with Variables

A *variable* is a symbol used to represent one or more numbers. A *variable expression* is a mathematical phrase that uses numbers, variables, and operation symbols. Variable expressions are made up of one or more terms. A *term* is a number, a variable, or the product or quotient of a number and a variable.

Example

The cost of renting a video from Fred's Movie Shed is $3/da. Write an equation you can use to find the total cost when you know how many days a video has been rented.

Relate   The total cost of the video rental is   3   times the number of days.

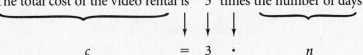

Write c $=$ 3 $\cdot$ n

$c = 3n$

Suppose Fred lowers the rental cost to $2/da. Write an equation to find the cost of renting a video for n days.

$c = 2n$

Practice

Write the next three numbers in each pattern.

1. 2, 4, 6, ▇, ▇, ▇   **2.** 3, 6, 9, ▇, ▇, ▇   **3.** 0, 4, 8, ▇, ▇, ▇

4. −3, −5, −7, ▇, ▇, ▇   **5.** 2, 4, 8, ▇, ▇, ▇   **6.** 17, 14, 11, ▇, ▇, ▇

Use an equation to model each situation.

7. A jet airliner travels 10 mi/min. Write an equation you can use to find the total distance traveled when you know how many minutes the jet has flown.

8. Write an equation you can use to find how many hours you must work at $4.75/h to earn $64.00.

Use an equation to model the relationship in each table.

9.

Days	Cost
1	$30
2	$60
3	$90
4	$120

10.

Minutes	Pages
10	40
20	80
30	120
40	160

11.

Weeks	Miles
2	300
4	500
6	900
8	1200

12.

Used	Left
0	15
3	12
9	6
15	0

© Prentice-Hall, Inc.

Lesson 1-3 Order of Operations

Evaluating Expressions

Order of Operations

1. Perform any operations inside grouping symbols.
2. Simplify any term with exponents.
3. Multiply and divide in order from left to right.
4. Add and subtract in order from left to right.

Example

Evaluate $6x + 4 - 2y^2$ for $x = 5$ and $y = 3$.

$$6x + 4 - 2y^2 = 6 \cdot 5 + 4 - 2 \cdot 3^2 \qquad \longleftarrow \textbf{Substitute 5 for } x \textbf{ and 3 for } y.$$

$$= 6 \cdot 5 + 4 - 2 \cdot 9 \qquad \longleftarrow \textbf{Simplify } 3^2.$$

$$= 30 + 4 - 18 \qquad \longleftarrow \textbf{Multiply first.}$$

$$= 16 \qquad \longleftarrow \textbf{Then add.}$$

Practice

Find the value of each expression.

1. $6 - 5 + 11$ **2.** $4 - 7 - 9$ **3.** $3 \cdot 2 \cdot 5$

4. $4 \div 4 \cdot 18$ **5.** $-7 + 6 - 8$ **6.** $9 \div 3 \cdot 4$

Simplify each numerical expression.

7. $4 + 12 \div 4$ **8.** $5 + 21 \div 7$ **9.** $8 \cdot 5 + 16 \div 4$ **10.** $11 - 12 \div 6$

11. $4^2 - 3 \cdot 5$ **12.** $4 \cdot 7^2$ **13.** $2 \cdot 5^2 - 8 \cdot 3$ **14.** $6^2 \div 9 - 10 \div 2$

Evaluate each expression.

15. $5 + x \cdot 3$ for $x = 2$ **16.** $4 \cdot 12 - g$ for $g = 24$ **17.** $7 + 5k \div 6$ for $k = 12$

18. $4z^2 - 42$ for $z = 3$ **19.** uv^2 for $u = 7$ and $v = 4$ **20.** $3m - n$ for $m = 2.1$ and $n = 1.7$

21. The largest of the Egyptian pyramids has a height *h* of about 137 m. One side of the pyramid's base has a length *l* of about 230 m. What is the volume of the pyramid? The formula for the volume of a pyramid is $V = \frac{1}{3}hl^2$.

22. The radius of an asteroid orbiting the Earth is 10 km. The asteroid can be thought of as a sphere. What is the volume of the asteroid. The formula for the volume of a sphere is $V = \frac{4}{3}\pi r^3$; use 3.14 for π.

© Prentice-Hall, Inc.

Lesson 1-3 Order of Operations

Evaluating Expressions With Grouping Symbols

When you evaluate expressions, work within the grouping symbols first. A fraction bar is a grouping symbol. Do any calculations above or below a fraction bar before simplifying the fraction.

Example

Evaluate $2\left(\dfrac{6 - 2x + 7}{3y - 5}\right)^2$ for $x = 4$ and $y = 2$.

$$2\left(\dfrac{6 - 2x + 7}{3y - 5}\right)^2 = 2\left(\dfrac{6 - 2 \cdot 4 + 7}{3 \cdot 2 - 5}\right)^2 \quad \longleftarrow \text{ Substitute 4 for } x \text{ and 2 for } y.$$

$$= 2\left(\dfrac{6 - 8 + 7}{6 - 5}\right)^2 \quad \longleftarrow \text{ Perform multiplication in numerator and denominator.}$$

$$= 2\left(\dfrac{5}{1}\right)^2 \quad \longleftarrow \text{ Perform addition and subtraction.}$$

$$= 2(5)^2 \quad \longleftarrow \text{ Simplify fraction.}$$

$$= 2(25) \quad \longleftarrow \text{ Square number in parentheses.}$$

$$= 50$$

Practice

Find the value of each expression.

1. $(12 \div 3) + 10$ **2.** $(3 \cdot 4 \cdot 5) + 7$ **3.** $(5 + 11) \div 28$

4. $(5 \cdot 6) - 14$ **5.** $7 \cdot (5 + 4)$ **6.** $(0.7 \cdot 46) + 5.4$

Simplify each numerical expression.

7. $(7 - 4) \cdot 10$ **8.** $\dfrac{3(4 - 2)}{6}$ **9.** $6(18 \div 9 + 4)$ **10.** $3(6 + 1) - 1$

11. $5 - 2(4 + 7)$ **12.** $\dfrac{(7 - 4)^2}{8}$ **13.** $7 - [4(18 \div 6) + 1]$ **14.** $8\left(\dfrac{3 + 1}{32 - 16}\right)$

Evaluate each expression.

15. $\dfrac{p + 5}{4}$ for $p = 11$ **16.** $(4m)^2$ for $m = 2$ **17.** $3(2 - 8b) - 7$ for $b = -1$

18. $2\left(\dfrac{6x - 9}{3}\right)$ for $x = 4$ **19.** $c[4 + 2(c - 5)]$ for $c = 3$ **20.** $\dfrac{t^2}{11 - t}$ for $t = 5$

21. The Plains Indians of the American Midwest lived in portable, cone-shaped houses called tepees. A typical tepee had a height h of 5 m. The radius of the tepee's base was 4 m. What was the volume of the tepee? The formula for the volume of a cone is $V = \dfrac{\pi r^2 h}{3}$.

22. Suzy Allen McKay wants to wallpaper her kitchen with two types of wallpaper. She requires 10 ft of wallpaper X and 5 ft of wallpaper Y, where x is the number of feet for wallpaper X and y is the number of feet for wallpaper Y. The cost to wallpaper Mrs. McKay's kitchen can be found using the formula $C = 2x^2 + 4y^2$. What is the cost to wallpaper Mrs. McKay's kitchen?

Lesson 1-4 Adding and Subtracting Integers

Part 1

Adding Integers and Decimals

Example

Your aunt asks you to go to the store for her. She gives you $8. You go to the store and buy $6 worth of merchandise for her. How much money do you have left?

Method 1: Use a number line.

Start at 0. Move right 8 units.   Then move left 6 units.

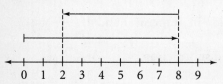

The result is 2.

$8 + (-6) = 2$

Method 2: Use tiles.

Start with 8 positive tiles and 6 negative tiles.

Make zero pairs.

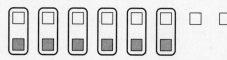

There are 2 positive tiles left.

$8 + (-6) = 2$

The result of the shopping trip is that you have $2 left.

Practice

Use a number line to find the value of each of the following.

1. Start at 4, increase 5.   **2.** Start at -2, increase 6.   **3.** Start at 4, decrease 8.

4. Start at -3, decrease 3.   **5.** Start at -4, increase 2.   **6.** Start at 5, decrease 5.

Find each sum using tiles.

7. $-7 + 3$ **8.** $2 + (-5)$ **9.** $-1 + (-4)$ **10.** $4 + (-1)$

11. $6 + (-4)$ **12.** $8 + (-3)$ **13.** $(-2) + (-7)$ **14.** $-9 + 9$

Solve each problem using a number line.

15. The temperature at 7 A.M. was $-5°F$. By noon the temperature had increased by $18°F$. At 7 P.M. the temperature had decreased by $26°F$. What is the temperature at 7 P.M.?

16. Antifreeze is added to water in a car radiator to prevent the water from freezing. Water freezes at $32°F$. If a certain mixture of antifreeze and water has a new freezing point of $-5°F$, by how many degrees was the freezing point lowered?

© Prentice-Hall, Inc.

Lesson 1-4 Adding and Subtracting Integers

Subtracting Integers and Decimals

Example

Find each difference.

a. $2 - 5$

☐ ☐   ← **Start with 2 positive tiles.**

☐ ☐ ☐ ▦ ☐ ▦ ☐ ▦   ← **Add zero pairs until there are 5 positive tiles.**

    ▦   ▦   ▦   ← **Remove 5 positive tiles.**

There are 3 negative tiles left.

$2 - 5 = -3$

b. $1 - (-4)$

☐   ← **Start with 1 positive tile.**

☐ ☐ ▦ ☐ ▦ ☐ ▦ ☐ ▦   ← **Add zero pairs until there are 4 negative tiles.**

☐ ☐ ☐   ☐   ☐   ← **Remove 4 negative tiles.**

There are 5 positive tiles left.

$1 - (-4) = 5$

Practice

Find the value of each expression.

1. $\left|-8\right|$ **2.** $\left|7\right|$ **3.** $\left|2 + 17\right|$

4. $\left|6 - 3\right|$ **5.** $\left|-3 + (-11)\right|$ **6.** $\left|3.8 - 2.6\right|$

Use tiles to find each difference.

7. $5 - 7$ **8.** $-3 - 4$ **9.** $-2 - (-4)$ **10.** $6 - 2$

11. $4 - 6$ **12.** $-1 - (-8)$ **13.** $-5 - 2$ **14.** $4 - (-3)$

15. $-1 - 9$ **16.** $3 - (-7)$ **17.** $8 - (-8)$ **18.** $4 - (-9)$

19. $1 - (-6)$ **20.** $-4 - (-7)$ **21.** $-3 - 9$ **22.** $9 - (-9)$

23. $-6 - (-2)$ **24.** $9 - 9$ **25.** $-2 - (-3)$ **26.** $4 - (-2)$

27. $8 - 8$ **28.** $-3 - (-4)$ **29.** $-1 - 8$ **30.** $2 - (-5)$

Lesson 1-5 Multiplying and Dividing Integers

Multiplying and Dividing

Part **1**

Multiplication and Division Rules

- The product or quotient of two positive numbers is positive.
- The product or quotient of two negative numbers is positive.
- The product or quotient of a positive number and a negative number is negative.

Example

Evaluate $-5p - \frac{6q}{r}$ for $p = -4$, $q = 3$, and $r = -9$.

$$-5p - \frac{6q}{r} = -5(-4) - \frac{6(3)}{-9}$$ ← **Substitute -4 for p, 3 for q, and -9 for r.**

$$= 20 - \frac{18}{-9}$$ ← **Multiply.**

$$= 20 - (-2)$$ ← **Divide.**

$$= 20 + 2$$ ← **The opposite of -2 is 2.**

$$= 22$$ ← **Add.**

Practice

Simplify each expression.

1. $1 + (-2) + 3$ **2.** $(-3) + (-4) + (-5)$ **3.** $4 + 6 + 8 + (-10)$

4. $(-12) + (-10) + (-7)$ **5.** $35 + (-34) + (-9) + 4$ **6.** $(-17) + 1 + (-9) + 12$

Simplify each expression.

7. $5(-6)$ **8.** $\frac{-32}{8}$ **9.** $\frac{(3-9)}{-3}$ **10.** $(-4)(-3)$

Evaluate each expression for $a = -3$, $b = -1$, and $c = 4$.

11. ac **12.** $\frac{b}{c}$ **13.** $2a - b$ **14.** $c - 5b$

15. A clothing store opened for business six months ago. The store lost money for three months before showing a profit. Find the mean of the store's monthly income: $-\$500$, $-\$350$, $-\$150$, $\$110$, $\$400$, $\$850$.

16. At the beginning of a long drought, a lake is 550 ft deep at its deepest point. During the drought, the change in lake depth is -4.5 ft/wk.

    a. Write an equation to find the lake depth w weeks after the beginning of the drought.

    b. How far has the lake level dropped after 13 wk?

    c. What is the lake depth after 13 wk?

© Prentice-Hall, Inc.

Lesson 1-5 Multiplying and Dividing Integers

Part 2

Simplifying Expressions with Exponents

Order of Operations

1. Perform any operations inside grouping symbols.
2. Simplify any term with exponents.
3. Multiply and divide in order from left to right.
4. Add and subtract in order from left to right.

Example

Use the order of operations to simplify expressions.

$$-2^4 = -(2 \cdot 2 \cdot 2 \cdot 2)$$

$$= -16$$

$$(-2)^4 = (-2 \cdot -2 \cdot -2 \cdot -2)$$

$$= 16$$

Practice

Simplify.

1. 3^3 **2.** 4^2 **3.** 2^5 **4.** 5^3

Simplify.

5. $(-6)^3$ **6.** $-(-2)^4$ **7.** $-(3)^3$ **8.** -7^2 **9.** $(-3)^3$

10. -3^3 **11.** -2^4 **12.** $(-2)^4$ **13.** $-(-5)^2$ **14.** $-(-2)^5$

15. $-1(-1)^7$ **16.** $(-7)^2$ **17.** -5^3 **18.** -5^2 **19.** $(-4)^3$

Evaluate each expression for $r = 2$, $s = 3$, and $t = 4$.

20. $(-r)^s$ **21.** $-t^r$ **22.** r^r **23.** $(-5)^t$ **24.** 6^r

25. $(-t)^s$ **26.** -3^t **27.** s^t **28.** $(r \cdot s)^r$ **29.** $(-s)^t$

30. -4^r **31.** 3^s **32.** s^2 **33.** t^3 **34.** $-r^s$

35. $(s + t)^r$ **36.** $(-t)^s$ **37.** -5^s **38.** $-s^s$ **39.** -7^t

Lesson 1-6 Real and Rational Numbers

Comparing Rational Numbers

When you compare two real numbers, only one of these can be true:

$a < b$ or $a = b$ or $a > b$

less than   equal to   greater than

Example

The two fractions $-\frac{7}{11}$ and $-\frac{3}{7}$ can be compared with a number line or with a calculator.

On a number line:

$$-1 \quad -\frac{7}{11} \quad -\frac{3}{7} \quad 0 \quad \frac{3}{7} \quad \frac{7}{11} \quad 1$$

⟵ **Numbers are greater as you move to the right on a number line.**

With a calculator:

$-\frac{7}{11}$ [(-)]   7   [÷]   11   [ENTER] -0.6363636

$-\frac{3}{7}$ [(-)]   3   [÷]   8   [ENTER] -0.2727272

$-0.\overline{63} < -0.\overline{27}$

Using either method, the two fractions can be written as the inequality

$-\frac{7}{11} < -\frac{3}{7}$.

Practice

Use $<$, $=$, or $>$ to compare.

1. -4 ■ -3 **2.** -8 ■ 2 **3.** 1 ■ -6

4. $\frac{15}{12}$ ■ $\frac{5}{4}$ **5.** $2\frac{5}{7}$ ■ $\frac{19}{7}$ **6.** -21 ■ -21.5

Write each group of numbers in order from least to greatest.

7. $-\frac{7}{12}, -\frac{8}{15}, -\frac{3}{4}$ **8.** $\frac{3}{8}, \frac{5}{12}, 0.43$ **9.** $-\frac{6}{13}, -\frac{2}{4}, -\frac{13}{25}$

10. $-0.37, -0.037, -0.35$ **11.** $\frac{2}{12}, -\frac{3}{11}, -\frac{3}{13}$ **12.** $\frac{10}{11}, \frac{9}{10}, \frac{11}{12}$

13. $-0.5, -0.55, -\frac{7}{13}$ **14.** $-\frac{7}{9}, -\frac{3}{5}, -\frac{5}{7}$ **15.** $-\frac{1}{2}, -\frac{1}{4}, -\frac{1}{3}$

16. $-0.125, -\frac{1}{7}, -\frac{1}{9}$ **17.** $-\frac{2}{3}, -\frac{4}{5}, -\frac{9}{16}$ **18.** $-0.45, -0.0045, -4.5$

19. $-0.45, -0.96, 67$ **20.** $\frac{9}{11}, \frac{10}{12}, \frac{9}{10}$ **21.** $\frac{10}{9}, \frac{11}{9}, -\frac{12}{9}$

22. $-\frac{10}{50}, -\frac{9}{18}, -\frac{3}{2}$ **23.** $\frac{36}{40}, \frac{1}{3}, \frac{10}{13}$ **24.** $-0.537, -0.648, -0.058$

Lesson 1-6 Real Numbers and Rational Numbers

Part **2**

Evaluating Expressions

You evaluate expressions using rational numbers by substituting and performing the indicated operations.

Example

Evaluate $3g + 2h$ where $g = \frac{3}{7}$ and $h = -\frac{2}{3}$.

$$3g + 2h = 3\left(\frac{3}{7}\right) + 2\left(-\frac{2}{3}\right) \quad \longleftarrow \textbf{Substitute the values for } g \textbf{ and } h.$$

$$= \frac{9}{7} - \frac{4}{3}$$

$$= -\frac{1}{21}$$

Practice

Evaluate.

1. $3x$ for $x = 2$ **2.** $a + 2$ for $a = 6$

3. $s + t$ for $s = 8, t = 4$ **4.** $3(f - 4)$ for $f = 2$

5. $2kl$ for $k = 5, l = 6$ **6.** $3d - e$ for $d = 8, e = 4$

Evaluate.

7. $2x + \frac{5}{4}$ for $x = \frac{2}{3}$ **8.** $\frac{a}{4} + \frac{2}{b}$ for $a = 6, b = 3$ **9.** $\frac{3}{4}(f - 4)$ for $f = 2\frac{1}{3}$

10. $3x + 5$ for $x = \frac{1}{3}$ **11.** $2x + \frac{1}{y}$ for $x = 4, y = 3$ **12.** $3h + r$ for $h = \frac{3}{4}, r = 4$

13. $(x - x)y$ for $x = 2, y = \frac{1}{2}$ **14.** $\frac{1}{3}(g + h)$ for $g = 7, h = 8$ **15.** qrt for $q = \frac{1}{2}, r = \frac{1}{3}, t = \frac{1}{4}$

16. $\frac{rt}{q}$ for $r = 2, t = 4, q = 7$ **17.** $\frac{(x + y)}{6}$ for $x = 1, y = 11$ **18.** $\frac{c}{5} + \frac{2}{d}$ for $c = 4, d = 6$

19. Use the expression $\frac{s}{10}$ to find the equivalent inches of rainfall from the number of inches of snow where s is the number of inches of snow.

    a. What is the equivalent amount of rain from 14 in. of snow?

    b. What is the equivalent amount of rain from $6\frac{1}{2}$ in. of snow?

    c. What is the equivalent amount of rain from $15\frac{1}{2}$ in. of snow?

20. Use the expression $\frac{8m}{5}$ to find the equivalent number of kilometers where m is the number of miles.

    a. What is the equivalent number of kilometers from 5 mi?

    b. What is the equivalent number of kilometers from $3\frac{1}{2}$ mi?

    c. What is the equivalent number of kilometers from $\frac{1}{3}$ mi?

Lesson 1-7 Experimental Probability and Simulation Part 1

Finding Experimental Probability

Probability measures how likely an event is to happen. Experimental probability is a ratio based on data gathered through observation or experimentation. Use this ratio to find experimental probability.

$$P(\text{event}) = \frac{\text{number of times an event happens}}{\text{number of times an experiment is done}}$$

Example

Suppose you have a box of tennis balls. You choose ten tennis balls at random and find that four of them are hot pink. What is the $P(\text{hot pink})$?

$$P(\text{hot pink}) = \frac{\text{number of hot pink tennis balls}}{\text{number of tennis balls selected}}$$

$$= \frac{4}{10} = \frac{2}{5} = 0.4 = 40\%$$

The probability that a tennis ball is hot pink is 40%.

Practice

Express each fraction as a percent. Round to the nearest tenth.

1. $\frac{1}{2}$ 2. $\frac{12}{25}$ 3. $\frac{15}{23}$

4. $\frac{38}{100}$ 5. $\frac{6}{7}$ 6. $\frac{79}{87}$

Suppose you select a shell at random from a box of shells of the type and number in the table. Find each probability as a percent, rounded to the nearest tenth.

7. Find $P(\text{moon snail})$

8. Find $P(\text{coquina})$

9. Find $P(\text{periwinkle})$

10. Find $P(\text{razor clam})$

Shells Collected

atlantic auger	12
periwinkle	13
moon snail	3
lettered olive	5
razor clam	16
coquina	45
Total	**94**

Suppose you select a marble at random from a bowl of marbles of the color and number in the table. Find each probability as a fraction; and as a percent, rounded to the nearest tenth.

11. Find P (blue marble)

12. Find P (white marble)

13. Find P (orange marble)

14. Find P (yellow marble)

Bowl of Marbles

blue	15
yellow	9
orange	11
green	2
white	4
clear	22
Total	**63**

Lesson 1-7 Experimental Probability and Simulation

Part 2

Conducting a Simulation

A simulation is a model of a real-life situation. Simulations are performed using randomly generated numbers to approximate the actual situation.

Example

You perform a simulation to predict the probabiltiy of a red traffic light. To simulate the color of the light, you use a number cube. You assign numbers 1, 2, and 3 to a green light, numbers 4 and 5 to a red light, and number 6 to represent a yellow light. Your rolls are in the chart below.

4	2	6	6	3	5	1	2	5	3	6	3	2	5	2	1	5	6	3	4

$$P(\text{red light}) = \frac{\text{number of times an event happens}}{\text{number of times the experiment is done}} = \frac{6}{20}$$

$$= \frac{3}{10}$$

$$= 0.3 \text{ or } 30\%$$

Practice

Express each fraction as a percent. Round to the nearest tenth.

1. $\frac{3}{15}$ **2.** $\frac{2}{12}$ **3.** $\frac{12}{20}$

4. $\frac{3}{5}$ **5.** $\frac{9}{14}$ **6.** $\frac{1}{6}$

Suppose you breed golden retrievers. You want to simulate the number of male and female puppies in a litter. Because each puppy in the litter has an equal chance of being male or female, toss a coin to determine the sex of each puppy. Assume the golden retriever has 12 puppies. Create a data table and determine each probability.

7. $P(\text{male})$ **8.** $P(\text{female})$

Repeat your simulation for a litter of 14 puppies.

9. What is $P(\text{male})$?   **10.** What is $P(\text{female})$?

Suppose you want to predict the outcome of your favorite soccer team. One of three outcomes is possible: win, lose, or tie—all with the same chance of happening. To simulate whether your team wins, loses, or ties, use a numbered cube. Assign numbers 1 and 2 to a win, 3 and 4 for a loss, and 5 and 6 for a tie. Roll the number cube 10 times to simulate 10 games. Record the ten-game season in a data table.

11. $P(\text{win})$ **12.** $P(\text{loss})$ **13.** $P(\text{tie})$

Lesson 1-8 Organizing Data in Matrices

A matrix is a rectangular arrangement of numbers. The numbers are arranged in rows and columns and are usually written inside brackets. You identify the size of a matrix by the number of rows and the number of columns. You can add and subtract matrices only if they are the same size. You do this by adding or subtracting the corresponding entries.

Example

corresponding entries   solution

$$\begin{bmatrix} 5 & 7 \\ 8 & 4 \end{bmatrix} + \begin{bmatrix} 3 & 6 \\ 2 & 1 \end{bmatrix} = \begin{bmatrix} 8 & 13 \\ 10 & 5 \end{bmatrix}$$

corresponding entries   solution

$$\begin{bmatrix} 5 & 7 \\ 8 & 4 \end{bmatrix} - \begin{bmatrix} 3 & 6 \\ 2 & 1 \end{bmatrix} = \begin{bmatrix} 2 & 1 \\ 6 & 3 \end{bmatrix}$$

Practice

 Simplify.

1. $-22 + (-15)$ **2.** $13 + (-8)$ **3.** $4 - (-8)$

4. $-27 + 27$ **5.** $14 - 23$ **6.** $-16 + 30$

Add each pair of matrices. Then subtract the second matrix from the first matrix in each pair.

7. $\begin{bmatrix} 4 & 8 \\ 10 & 9 \end{bmatrix}, \begin{bmatrix} 5 & 2 \\ 8 & 3 \end{bmatrix}$ **8.** $\begin{bmatrix} 5 & 11 & 9 \\ 4 & 10 & 7 \end{bmatrix}, \begin{bmatrix} 2 & 6 & 9 \\ 1 & 8 & 4 \end{bmatrix}$ **9.** $\begin{bmatrix} 5 & 2 \\ 8 & 7 \\ 1 & 9 \end{bmatrix}, \begin{bmatrix} 9 & 3 \\ 2 & 4 \\ 5 & 0 \end{bmatrix}$

10. $\begin{bmatrix} 1 & 0 \\ 0 & 1 \end{bmatrix}, \begin{bmatrix} 2 & 5 \\ 7 & 9 \end{bmatrix}$ **11.** $\begin{bmatrix} 2 & 3 & 1 \\ 1 & 3 & 6 \\ 7 & 2 & 4 \end{bmatrix}, \begin{bmatrix} 1 & 5 & 4 \\ 2 & 6 & 7 \\ 9 & 1 & 2 \end{bmatrix}$ **12.** $\begin{bmatrix} 2 & 1 & 5 \\ 6 & 0 & 9 \end{bmatrix}, \begin{bmatrix} 8 & 8 & 8 \\ 0 & 5 & 1 \end{bmatrix}$

13. Write each table as a matrix. Then add the matrices to find the total shrimp catch for the two shrimp boats.

**Shrimp boat *Gulf Pride*
(catch in pounds)**

Day of Week	brown shrimp	white shrimp	pink shrimp
Wed	382	78	64
Thur	297	94	72
Fri	245	91	90

**Shrimp boat *Marie*
(catch in pounds)**

Day of Week	brown shrimp	white shrimp	pink shrimp
Wed	178	61	54
Thur	423	86	78
Fri	302	69	61

© Prentice-Hall, Inc.

Lesson 1-9 Variables and Formulas in Spreadsheets

Like a matrix, a spreadsheet organizes data in rows and columns. Each individual box is called a cell. A computer spreadsheet program uses different signs for some operations. For multiplication, spreadsheets use *. For division, spreadsheets use /, and to indicate "raise to a power," they use ^.

Example

This spreadsheet is used by a company to determine the paychecks for each employee. The spreadsheet contains each employee's name, rate of pay per hour, hours worked, and total pay.

Pay is calculated by multiplying the rate times the hours worked. The formula to calculate Misha's pay is B2*C2. The formula to calculate Rae's pay is B3*C3, and the formula to calculate Shawn's pay is B4*C4.

	A	B	C	D
1	Employee	Rate	Hours	Pay
2	Misha	4.75	23	109.25
3	Rae	5.25	19	99.75
4	Shawn	4.90	24	117.60

Practice

Evaluate each expression.

1. 12*10

2. 25/12.5

3. 3^3

4. (5*6)/2

5. 2^3/2^2

6. 3.5*4

Use the spreadsheet.

	A	B	C	D
1	x	$4x - 5$	$x^2 + 1$	$x - 6$
2	9			
3	3			

7. Write the formulas you would use in cells B2, C2, and D2 to evaluate the expressions in cells B1, C1, and D1.

8. Find the values for cells B2, C2, and D2.

9. Find the values for cells B3, C3, and D3.

© Prentice-Hall, Inc.

Lesson 2-1 Analyzing Data Using Scatter Plots

Drawing and Interpreting Scatter Plots

Example

Kurt wanted to find the relation between the length of a CD to the number of songs on a CD. He randomly chose seven CDs from his collection and made the following data table. Draw a scatter plot of the data.

Number of songs	CD length (min)
11	41
22	63
12	57
16	75
10	40
14	54
12	41

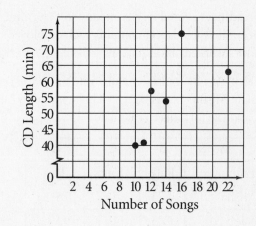

Practice

Graph each ordered pair.

1. (2, 5)   2. (3, 7)   3. (5, 0)

4. (4, 1)   5. (0, 6)   6. (8, 7.5)

7. Each ordered pair represents (daily low temperature, daily high temperature) for a week in Austin, Texas in January. Temperatures are in degrees Fahrenheit. Draw a scatter plot of the data.
(27, 55), (38, 68), (40, 71), (55, 73), (45, 72), (47, 77), (53, 77)

A group of students was interested in whether the weights of books relate to their prices. The students constructed a scatter plot of the data they collected.

8. About how much did the most expensive book cost?

9. About how much did the most expensive book weigh?

10. Was the most expensive book also the heaviest book?

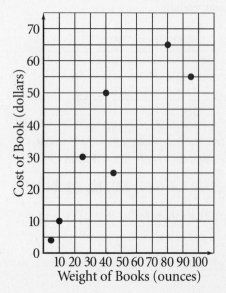

© Prentice-Hall, Inc.

Lesson 2-1 Analyzing Data Using Scatter Plots

Part **2**

Analyzing Trends in Data

Example

The scatter plot at right shows information about the body masses of ten different species of birds and the numbers of eggs they typically lay in their nests. Is there a *positive correlation*, a *negative correlation*, or *no correlation* between the body masses and the number of eggs laid? Explain.

There is no correlation; the points are scattered about randomly.

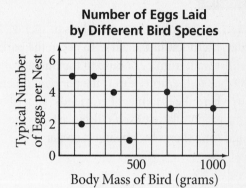

Number of Eggs Laid by Different Bird Species

Practice

Mini ▶ **Help**

Sketch a trend line through each scatter plot.

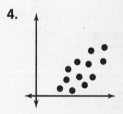

1.   2.   3.   4.

The records keeper for a high school football team made a scatter plot of total number of offensive yards gained and the total number of plays run in each game.

5. Draw a trend line on the scatter plot.

6. Is there a *positive correlation*, a *negative correlation*, or *no correlation* between the two data sets in the scatter plot?

7. Use the trend line to predict how many yards the offense will gain if they run 80 plays.

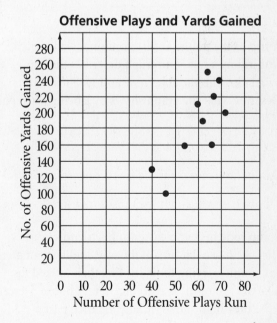

Offensive Plays and Yards Gained

© Prentice-Hall, Inc.

Lesson 2-2 Relating Graphs to Events

Part **1**

Interpreting Graphs

Example

Rosa and her friends have made plans to spend Saturday afternoon in the park. To get to the park, Rosa must walk from her house to the bus stop, ride the bus to the stop nearest the park, and then walk to where she is to meet her friends. Describe what the graph relating Rosa's distance from the park and time shows by labeling each part.

Going to the Park

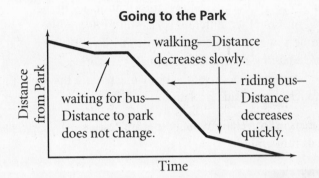

walking—Distance decreases slowly.

riding bus— Distance decreases quickly.

waiting for bus— Distance to park does not change.

Distance from Park

Time

Practice

Mini ► **Help**

Determine whether the line indicates that distance is increasing, decreasing, or staying the same.

1.   **2.**   **3.**   **4.**   **5.**

6. A trip to the public library combines walking with riding a bus. The graph describes the trip by relating the variables *time* and *total distance*. Describe what the graph shows by labeling each part.

A Trip to the Library

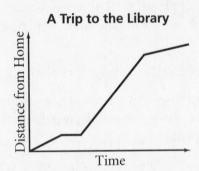

Distance from Home

Time

7. Philip rode his bicycle over to his friend's house. His bike had a flat tire on the way. Philip spent a few minutes trying to fix his tire, then walked his bike the rest of the way to his friend's house. Describe what the graph shows by labeling each part.

Philip's Bike Ride

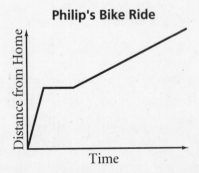

Distance from Home

Time

Lesson 2-2 Relating Graphs to Events

Sketching Graphs

When you draw a graph without actual data, the graph is called a sketch. A sketch is useful when you want to get an idea of what a graph looks like.

Example

Jill and Alejandro decided to go to dinner and a movie. Jill drove to Alejandro's house to pick him up. They went to a restaurant. After dinner, they went to the movie theater. After the movie, Jill dropped Alejandro off at his house before returning home. Sketch a graph to describe Jill's trip. Label the sections. Which is farther from Jill's house, Alejandro's house or the movie theater?

Alejandro's house is farther.

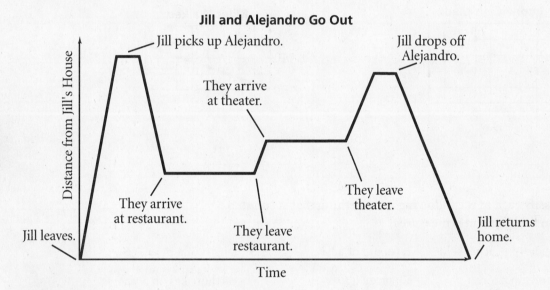

Jill and Alejandro Go Out

Practice

Sketch each graph relating distance to time.

1. increasing slowly
2. decreasing quickly
3. decreasing slowly
4. staying the same
5. increasing quickly

6. Sketch a graph to describe the movement of a basketball player running up and down the court during a basketball game.

7. You are in charge of planning a summer vacation for your family. Write a brief description of the places you will visit, what types of transportation you will use, and how long you will be gone. Sketch a graph to describe your vacation plan. Label the sections.

Lesson 2-2 Relating Graphs to Events

Classifying Data

Continuous data usually involves measurements, such as temperatures, lengths, and weight. *Discrete data* involves a count, like numbers of people or objects.

Example

Classify the data as *continuous* or *discrete*. Explain your reasoning and sketch a graph of each situation.

a. phone calls Tanya makes in one week
The number is discrete. Each day's count is distinct. The numbers of calls must be integers. There are no fractions of phone calls.

b. distance Tanya walks in one week
Distance is continuous. Every step Tanya walks adds an additional fraction of a mile to her total.

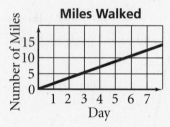

Practice

Classify each of the following as something that is counted or something that is measured.

1. cans of peaches sold   **2.** depth of a lake   **3.** length of a movie

4. trees in a forest   **5.** age of a fish   **6.** earthquakes in a year

Classify the data as *discrete* or *continuous*. Explain your reasoning and sketch a graph of each situation.

7. how long you sleep each night for one week   **8.** height of ocean between high and low tides

9. pressure in a tire as pumped with air   **10.** the position of the sun from 6 A.M. to 6 P.M.

11. the number of students at each football game for a season   **12.** species of birds visiting a feeder each month for a year

13. how much a newborn weighs during the first week   **14.** the number of players on a neighborhood baseball team each year

15. the size of a tree over a year   **16.** the number of movies watched each week

© Prentice-Hall, Inc.

Lesson 2-3 Linking Graphs to Tables

Example

Graph the data in the table on a coordinate plane.

Car Gasoline Consumption

Miles Driven	Gallons of Gasoline Consumed
5	0.2
10	0.3
15	0.5
20	0.7
25	0.8
30	1
35	1.2
40	1.3

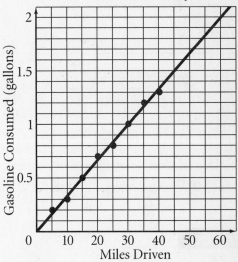

Gallons of gasoline consumed is the dependent variable.
Miles driven is the independent variable.

Practice

For each data set, draw a set of axes. Include the greatest and least values. Label all of the intervals.

1. (3, 5), (2, 8), (1, 15), (0, 17), (4, 3)

2. (3, 10), (−2, 2), (0, 4), (−1, 3), (2, 7)

3. (0.5, 5), (1.5, 15), (2, 20), (2.5, 25), (1, 10)

4. (40, −12), (25, 0), (10, 16), (35, −8), (30, −4)

Graph the data in each table.

5. Pizza Menu

Pizza Diameter	Price
10 in.	$4
12 in.	$5
16 in.	$7
20 in.	$11
24 in.	$16

6. Speed of Jet Aircraft

Miles per hour	Mach number
400	0.5
600	0.8
750	1.0
1000	1.4
1500	2.0

7. Bacterial Reproduction

Time	No. of Bacteria
0 hr	1
0.5 hr	3
1 hr	8
1.5 hr	23
2 hr	64

Lessson 2-4 Functions
Identifying Relations and Functions

Example

Determine if each relation is a function.

a.

x	y		
−2	5	→	(−2, 5)
−1	3	→	(−1, 3)
0	−7	→	(0, −7)
−1	5	→	(−1, 5)
2	−3	→	(2, −6)

This relation is not a function because two *y*-values, 3 and 5, are assigned to one *x*-value, −1.

b.

x	y		
2	2	→	(2, 2)
5	2	→	(5, 3)
7	2	→	(7, 2)
11	2	→	(11, 5)
14	2	→	(14, 7)

This relation is a function because exactly one *y*-value is assigned to each *x*-value.

Practice

Write the data in each table as a set of ordered pairs.

1.

x	y
0	4
−5	0
13	6
7	−11

2.

x	y
20	24
33	−15
57	14
19	8

3.

x	y
0.50	17
0.75	11
0.25	13
1.15	9

4.

x	y
76	115
0	−70
113	−12
−84	0

Determine if each relation is a function.

5.

x	y
1	3
3	1
4	6
6	4

6.

x	y
12	0.6
11	−0.15
11	0.15
10	−0.8

7.

x	y
−3	7
−5	4
−7	4
−13	7

8.

x	y
82	30
93	25
96	25
82	25

Which relations in Exercises 9–12 are functions? If the relation is a function, identify the dependent and independent variables.

9. students' shoe sizes and eye colors

10. ocean depth and pressure

11. length of movie and price of admission

12. test grade and number of correct answers

Lesson 2-4 Functions

Part **2**

Evaluating Functions

A *function rule* is an equation that describes a function. If you know the input values, you can use a function rule to find the output values. The *domain* is the set of all possible input values. The *range* of a function is the set of all possible output values.

Example

Evaluate the function rule $y = -3x^2 + 2$ for $x = 5$.

$$y = -3x^2 + 2$$

$$y = -3(5)^2 + 2 \quad \longleftarrow \textbf{Substitute 5 for } x.$$

$$y = -3(25) + 2 \quad \longleftarrow \textbf{Find } (5)^2 \textbf{ first.}$$

$$y = -75 + 2 \quad \longleftarrow \textbf{Then multiply.}$$

$$y = -73$$

When the x-value is 5, the y-value is -73.

Practice

 Evaluate the following expressions.

1. $3(2) + 1$ **2.** $-3(5) + 2$ **3.** $4 - (3)^2$

4. $-(4)^2 + 7$ **5.** $11 + (-6)^2$ **6.** $-2(1.5) + 0.5$

7. $9 - \frac{1}{4}(4)^2$ **8.** $-3(\frac{1}{2})^2 + \frac{1}{4}$ **9.** $(-7)^2 - 20$

10. $(3)^3 + 5$ **11.** $3(-2)^3 - 1$ **12.** $9(\frac{2}{3})^2 - 1$

Evaluate each function rule for $x = 4$.

13. $y = x - 1$ **14.** $y = 2x + 3$ **15.** $y = -x - 5$ **16.** $y = -7 + 3x$

17. $y = x^2 + 4$ **18.** $y = -3x^2 - 5$ **19.** $y = -\frac{3}{2}x + 5$ **20.** $y = 50 - x^3$

Find the range of each function when the domain is $\{-3, 0, 5\}$.

21. $y = 5x$ **22.** $y = 3x - 4$ **23.** $y = -x + 5$ **24.** $y = \frac{1}{2}x + \frac{5}{2}$

25. $y = 10 - x^2$ **26.** $y = \frac{1}{3}x^2 - 4$ **27.** $y = -x^3 - 8$ **28.** $y = x^2 + 1.4$

Find the range of each function when the domain is $\{-3, 0, 4\}$.

29. $y = 2x + 3$ **30.** $y = 3x - 4$ **31.** $y = -\frac{1}{4}x - 1$ **32.** $y = \frac{1}{3}x + 1$

33. $y = -2x + 4$ **34.** $y = 4x$ **35.** $y = -\frac{2}{3}x + \frac{1}{3}$ **36.** $y = -x^2 - 2.1$

Lesson 2-4 Functions
• **Part**
Analyzing Graphs   **3**

Example

Use the vertical line test to determine if each graph is the graph of a function.

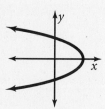

Because a vertical line (the *y*-axis in this example) passes through the graph more than once, this graph does not represent a function.

Because a vertical line passes through the graph only once, the graph represents a function.

Practice

  Identify each line as vertical, horizontal or neither.

1. 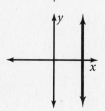   **2.**    **3.**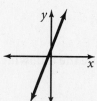

4. 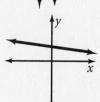   **5.** 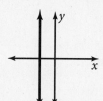   **6.**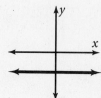

Use the vertical line test to determine if each graph represents a function.

7. 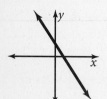   **8.** 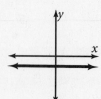   **9.** 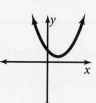   **10.**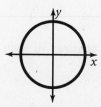

11.    **12.** 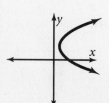   **13.** 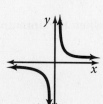   **14.**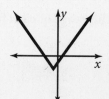

Lesson 2-5 Writing a Function Rule

Part 1

Understanding Function Notation

A function rule is an equation that describes a function. Sometimes the equation is written using *function notation*. To write a function rule in function notation, you use the symbol $f(x)$ in the place of y.

Example

Write $y = -3x^2 + 5$ in function notation.

$f(x) = -3x^2 + 5$.

Find the output value of the function when the input value is -2.

$f(-2) = -3(-2)^2 + 5$

$f(-2) = -3(4) + 5$

$f(-2) = -12 + 5$

$f(-2) = -7$

Practice

 Mini ▶ Help

Evaluate each expression.

1. $5x - 2$ for $x = 3$ **2.** $-12 + x$ for $x = -8$ **3.** $-4x + 7$ for $x = 6$

4. $-7 - x^2$ for $x = 5$ **5.** $\frac{3}{2}x + 4$ for $x = -4$ **6.** $3x^2 - 1$ for $x = \frac{2}{3}$

Use function notation to write an equivalent equation for each of the following.

7. $y = -2x + 5$ **8.** $y = 7 - x^2$ **9.** $y = 0.3x + 1.6$ **10.** $y = 4x^3 - 57$

Find $f(5)$ for each function.

11. $f(x) = x + 2$ **12.** $f(x) = 4x - 4$ **13.** $f(x) = 17 - x^2$ **14.** $f(x) = -11x + 33$

15. $f(x) = \frac{4}{5}x - 6$ **16.** $f(x) = 2 + 3x$ **17.** $f(x) = 1.4x - 7$ **18.** $f(x) = 3 + x^2$

Find the range of each function when the domain is $\{-2, 1, 7\}$.

19. $f(x) = 5 - 4x$ **20.** $f(x) = 3x + 2$ **21.** $f(x) = 2x^2 + 1$ **22.** $f(x) = -\frac{1}{2}x - 8$

23. $f(x) = 4 + x$ **24.** $f(x) = 6x - 7$ **25.** $f(x) = -7x^2$ **26.** $f(x) = 2.4x + 6.9$

Find the range of each function when the domain is $\{-1, 0, 2\}$.

27. $f(x) = 10x + 1$ **28.** $f(x) = x^2 - 1$ **29.** $f(x) = 4x - 2$ **30.** $f(x) = 4 - 3x$

31. $f(x) = 3x^2 + 1$ **32.** $f(x) = 1.3x + 4.7$ **33.** $f(x) = -3x^2$ **34.** $f(x) = \frac{1}{2}x - 4$

Lesson 2-5 Writing a Function Rule

Using a Table of Values

Example

Write a function rule for each table.

a.

x	f(x)
−2	−6
−1	−3
1	3
3	9

Ask yourself, "What can I do to −2 to get −6, −1 to get −3, . . . ?"

$f(x)$ equals three times x

$f(x) = 3 \cdot x$

b.

x	f(x)
−3	7
−1	5
0	4
2	2

Ask yourself, "What can I do to −3 to get 7, −1 to get 5, 0 to get 4, . . . ?"

$f(x)$ equals 4 minus x

$f(x) = 4 - x$

Practice

Write two function rules for each pair of input and output values.

1. input: 1; output: 3   **2.** input: 2; output: −2   **3.** input: 5; output: 4

4. input: 3; output: 9   **5.** input: 6; output: −5   **6.** input: 7; output: 13

Match each table of values with a given rule on the right.

7.

x	f(x)
−2	−10
0	0
1	5
2	10

8.

x	f(x)
3	0
5	−2
7	−4
9	−6

9.

x	f(x)
−2	−1
0	0
4	2
6	3

A. $f(x) = 3 - x$

B. $f(x) = \frac{1}{2}x$

C. $f(x) = 5x$

Write a function rule for each table.

10.

x	f(x)
−3	0
0	3
2	5
5	8

11.

x	f(x)
−2	−7
−1	−6
3	−2
5	0

12.

x	f(x)
−3	−1
0	0
3	1
6	2

13.

x	f(x)
−0.5	1
1	−2
−3	6
2	−4

Lesson 2-5 Writing a Function Rule

Using Words to Write a Rule

Example

Tap water must be treated with a chemical conditioner before it is safe for the fish in an aquarium. The directions on the water conditioner say to add three drops conditioner to every gal of water. Write a function rule to describe this relationship.

Define g = gallons of water

$D(g)$ = drops of conditioner needed

Relate   drops of conditioner needed   is   three   times   gallons of water

Write $D(g)$ =   3   · g

The rule $D(g) = 3g$ describes the relationship.

Practice

Identify the independent and dependent variables for each situation.

1. cost of filling a car with gasoline, number of gallons required

2. height of ocean tide, position of moon

3. amount you deposit in your savings account, balance of your account

For Exercises 4–6, write a function rule to describe each statement.

4. change from a five-dollar bill when buying stamps for 32¢ each

5. income from recycling aluminum cans for 5¢ per can

6. profit from typing papers for $1 per page after investing $500 in a computer

7. The kilometer is the unit used to measure long distances in most countries. There are about 1.6 km in every mile.

   a. Write a function rule relating miles to kilometers.

   b. How many kilometers are in 10 mi?

8. Suppose you decide to make some extra money by raking leaves. You spend $25.00 on a rake and other supplies. You charge $1.50 per bag of leaves.

   a. Write a rule to describe your profit as a function of the number of bags of leaves you rake.

   b. How many bags of leaves will you have to rake to equal your initial investment?

Lesson 2-6 The Three Views of a Function

Example

You hear on the radio that a thunderstorm is moving toward your home at 15 mi/h. At noon the thunderstorm is 60 mi away. The distance of the storm from your home $D(t)$ is a function of the time that has elapsed since noon. Use the rule $D(t) = 60 - 15t$ to make a table of values and then graph.

Step 1:
Choose values for t that seem reasonable, like 1, 2, 3, 5.

Step 2:
Input the values for t. Evaluate to find $D(t)$.

t	$D(t) = 60 - 15t$	$(t, D(t))$
1	$60 - 15(1) = 45$	$(1, 45)$
2	$60 - 15(2) = 30$	$(2, 30)$
3	$60 - 15(3) = 15$	$(3, 15)$
5	$60 - 15(5) = -15$	$(5, -15)$

Step 3:
Plot the ordered pairs.

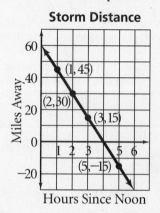

Storm Distance

When will the storm be directly over your home?

It will arrive when $D(t) = 0$, which corresponds to $t = 4$.

What do negative values of $D(t)$ indicate?

They indicate that the storm has passed over you and is moving away.

Practice

Find $f(5)$ and $f(-3)$ for each function.

1. $f(x) = 8 - x$

2. $f(x) = 6x + 2$

3. $f(x) = 5 - 3x$

4. $f(x) = 7 - |2x|$

5. $f(x) = |x - 4|$

6. $f(x) = |3x - 5|$

7. At 2 P.M. a train is 330 mi from a station. The train is traveling toward the station at 55 mi/h.

    a. Use the rule $D(t) = 330 - 55t$ to make a table of values.

    b. Graph the function.   **c.** At what time will the train arrive at the station?

Make a table of values for each graph.

8.

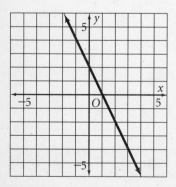

9.

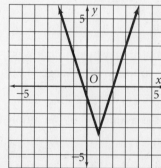

10.

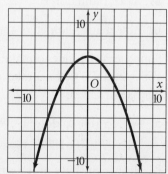

Lesson 2-7 Families of Functions

Part 1

Identifying the Family of an Equation

Example

To what family of functions does each equation belong? Explain.

a. $y = -5x + 2$

The highest power of x in the equation is 1. So, $y = -5x + 2$ is a *linear function*.

b. $y = 3x^2$

The highest power of x in this equation is 2. So, $y = 3x^2$ is a *quadratic function*.

c. $y = |4 - x|$

There is an absolute value symbol around the variable expressions. So $y = |4 - x|$ is an *absolute value function*.

Practice

Identify the highest power of x for each equation.

1. $y = x^2 + 1$

2. $y = 3x + 4$

3. $y = -x^2 + 6x - 3$

4. $y = \frac{7}{4}x$

5. $y = 11 + 6x - x^2$

6. $y = 2x^3 - x^2 + 5x$

To what family of functions does each equation belong? Explain.

7. $y = x^2 - 5$

8. $y = 3 + 4x$

9. $y = -2x - 17$

10. $y = |x + 1|$

11. $y = 3x^2 + 12$

12. $y = 7 - \frac{2}{3}x$

13. $y = 5 + 3|x|$

14. $y = x - 6x^2$

15. $y = -|2 - 4x|$

16. $y = 15 - 11x$

17. $y = x^2 + 2$

18. $y = 24x - 3 - x^2$

19. $y = 2x^2 + 6$

20. $y = 4|x| + 2$

21. $y = 3x - 4 + x^2$

22. $y = \frac{3}{4}x + 6$

23. $y = -\frac{3}{4}|x| - 3$

24. $y = x - 8$

25. $y = -x^2 + 3x$

26. $y = 4x + 2$

27. $y = 3 - x$

28. $y = |x + 7|$

29. $y = |x| + 5$

30. $y = 7 + x^2 - x$

31. $y = x^2 + 2$

32. $y = x + 4x^2$

33. $y = 3x - 2$

34. $y = 2x$

35. $y = 4 - |x|$

36. $y = |6 - x|$

37. $y = 2x^2 + 4x + 1$

38. $y = 6 - x$

39. $y = 1.4|x|$

40. Open-ended Create three equations for each family of functions: linear, quadratic, and absolute value.

Lesson 2-7 Families of Functions

Identifying the Family of a Graph

Example

To what family does each graph belong? Explain.

1.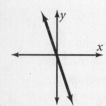

The graph is a straight line. So, it is a *linear function*.

2.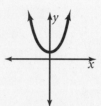

The graph is U-shaped. So, it is a *quadratic function*.

3.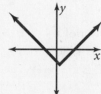

The graph forms a "V." So, it is an *absolute value function*.

Practice

Mini▶ **Help**

Describe each graph as a straight line, U-shaped, or V-shaped.

1.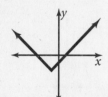

2.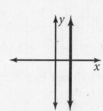

3.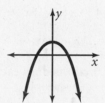

To what family of functions does each graph belong? Explain.

4.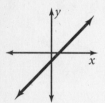

5.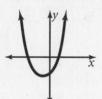

6.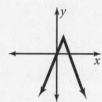

7.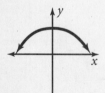

8.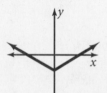

9.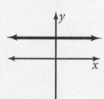

10.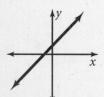

11.

12.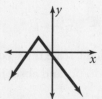

13. **Open-ended** Sketch three graphs that belong to the quadratic family of functions.

© Prentice-Hall, Inc.

Lesson 2-8 The Probability Formula

Finding Theoretical Probability

Example

Suppose you toss a coin three times. What is the probability of getting two heads and one tail?

There are eight possible outcomes for tossing a coin three times: HHH, HHT, HTH, THH, HTT, THT, TTH, TTT. The outcomes that result in two heads and one tail are HHT, HTH, and THH. So, there are three favorable outcomes.

$$P(\text{event}) = \frac{\text{number of favorable outcomes}}{\text{number of possible outcomes}}$$

$$P(\text{two heads and one tail}) = \frac{3}{8}$$

$$= 0.375 = 37.5\%$$

Practice

A parking lot contains eight sedans, five station wagons, four pickup trucks and three convertibles. What fraction of the cars does each represent?

1. station wagons

2. pickup trucks

3. not station wagons

4. sedans

5. not convertibles

6. sedans and pickup trucks

7. A multiple choice question has five possible responses, one correct and four incorrect.

    a. What is the number of possible outcomes?

    b. What is the probability of choosing a correct response?

    c. What is the probability of choosing an incorrect response if you guess randomly?

8. Suppose you have ten books placed randomly on a shelf. Three of the books are mysteries, five of the books are biographies, and two of the books are science fiction.

    a. What is the probability that the first book on the shelf is a mystery or a science fiction?

    b. What is the probability that the first book on the shelf is a romance?

    c. What is the probability that the last book on the shelf is not a mystery?

    d. You take a book from the shelf at random. What is the probability that the book is a biography?

Lesson 2-8 The Probability Formula

Using a Tree Diagram to Find a Sample Space

Example

Olivia is getting dressed for a party. She has a choice of two blouses (pink and yellow), two skirts (navy and gray), and two pairs of shoes (black and brown). Olivia is equally likely to choose any of the colors. What is the probability that she will wear a pink blouse and black shoes to the party?

Blouse	Skirt	Shoes	Sample Space	
pink	navy	black	pink, navy, black	← favorable outcomes:
		brown	pink, navy, brown	pink blouse and
	gray	black	pink, gray, black	← black shoes
		brown	pink, gray, brown	
yellow	navy	black	yellow, navy, black	
		brown	yellow, navy, brown	
	gray	black	yellow, gray, black	
		brown	yellow, gray, brown	

There are eight possible outcomes.

$$P(\text{pink blouse and black shoes}) = \frac{\text{number of favorable outcomes}}{\text{number of possible outcomes}} = \frac{2}{8} = \frac{1}{4}$$

The probability that Olivia will wear a pink blouse and black shoes to the party is $\frac{1}{4}$ or 25%.

Practice

Construct a factor tree for each number.

1. 25 **2.** 49 **3.** 16

4. 36 **5.** 60 **6.** 81

7. You work at a pizza restaurant. The restaurant sells two sizes of pizza (large and small), two types of crust (regular and deep-dish), and three toppings (pepperoni, mushrooms, and sausage).

   a. Use a tree diagram to show the sample space of one-topping pizzas.

   b. What is the probability that the next pizza you make will be deep-dish with mushrooms?

8. Erik, Tina, and Elisa are riding in the back seat of their parents' car.

   a. Use a tree diagram to show the sample space of seating order.

   b. What is the probability that Tina is sitting in the middle?

   c. What is the probability that Erik is sitting next to Elisa?

Lesson 3-1 Modeling and Solving Equations

Part 1

Solving Addition and Subtraction Equations

Example

Barbara wrote a check for $15. She calculated that this left a balance of $238 in her account. To find the amount in Barbara's account before she wrote this check, solve the equation $x - 15 = 238$.

$$x - 15 = 238$$
$$x - 15 + 15 = 238 + 15$$
$$x = 253$$

← **The inverse operation for subtraction is addition. Add 15 to each side.**

The amount in the account was $253.

Check
$$x - 15 = 238$$
$$253 - 15 \stackrel{?}{=} 238$$

← **Substitute 253 for x.**

$$238 = 238$$

Practice

Simplify.

1. $8.5 + 1.1$ 2. $-6 + 1$ 3. $-10 - 15$

4. $\frac{3}{8} + \frac{3}{4}$ 5. $15 - 6.2$ 6. $3 - 1\frac{1}{3}$

Solve and check.

7. $a - 11 = 5$ 8. $x + 2.5 = 7$ 9. $-3 + y = 15$

10. $r - 34 = 3$ 11. $n + \frac{1}{4} = \frac{3}{4}$ 12. $3 = 11 + q$

13. $p - 0.2 = 6.7$ 14. $-15 = z - 8$ 15. $t + 3.5 = 3.5$

Use the given equations to solve each problem.

16. The sum of the measures of two angles is 90°. One of these angles measures 73°. To find the measure of the other angle, solve the equation $m + 73 = 90$.

17. The thermometer read 16°F this morning. This is 20° higher than the reading yesterday morning. To find yesterday's reading, solve the equation $16 = t + 20$.

18. Sarah packs some books in a trunk. The total weight of the trunk with books is 65 lb. The empty trunk weighs 12 lb. To find the weight of the books, solve the equation $b + 12 = 65$.

© Prentice-Hall, Inc.

Lesson 3-1 Modeling and Solving Equations

Solving Multiplication and Division Equations

Part **2**

Example

$$\frac{x}{3} = -6.5 \qquad \longleftarrow \text{The operation is division. Multiply to undo.}$$

$$3\left(\frac{x}{3}\right) = 3(-6.5) \qquad \longleftarrow \text{Multiply each side of the equation by 3.}$$

$$x = -18.5$$

Check $\qquad \dfrac{x}{3} = -6.5$

$$\frac{-18.5}{3} \overset{?}{=} -6.5$$

$$-6.5 = -6.5$$

$$-48 = -10a \qquad \longleftarrow \text{The operation is multiplication. Divide to undo.}$$

$$\frac{-48}{-10} = \frac{-10a}{-10} \qquad \longleftarrow \text{Divide each side of the equation by } -10.$$

$$4.8 = a$$

Check $\qquad -48 = -10a$

$$-48 \overset{?}{=} -10(4.8)$$

$$-48 = -48$$

Practice

 Simplify.

1. $(-3)(1.8)$ ⁣⁣⁣ **2.** $\dfrac{-5.4}{-3}$ ⁣⁣⁣ **3.** $\left(\dfrac{2}{3}\right)\left(\dfrac{1}{2}\right)$

4. $(-1.5)(-2)$ ⁣⁣⁣ **5.** $1\frac{1}{2} + (-2)$ ⁣⁣⁣ **6.** $\dfrac{-6}{24}$

Solve and check.

7. $6n = -78$ ⁣⁣⁣ **8.** $18 = -2p$ ⁣⁣⁣ **9** $0.5t = 8.5$

10. $-\dfrac{a}{4} = -11$ ⁣⁣⁣ **11.** $\dfrac{r}{3} = -6.5$ ⁣⁣⁣ **12.** $\dfrac{z}{-18} = -18$

13. $12 = \dfrac{m}{-5}$ ⁣⁣⁣ **14.** $\dfrac{3}{4} = -2y$ ⁣⁣⁣ **15.** $3 = -\dfrac{x}{1}$

16. $10x = 0$ ⁣⁣⁣ **17.** $35 = 35p$ ⁣⁣⁣ **18.** $48 = 4.8w$

19. $-9t = 72$ ⁣⁣⁣ **20.** $-17t = 51$ ⁣⁣⁣ **21.** $8a = 56$

22. $\dfrac{1}{2}x = 2$ ⁣⁣⁣ **23.** $-\dfrac{2}{3}m = 4$ ⁣⁣⁣ **24.** $\dfrac{s}{5} = 11$

25. $\dfrac{2}{3}x = -\dfrac{9}{3}$ ⁣⁣⁣ **26.** $\dfrac{m}{4.5} = 33$ ⁣⁣⁣ **27.** $\dfrac{y}{9} = -5$

Lesson 3-1 Modeling and Solving Equations

Modeling by Writing Equations

Part **3**

Example

In 1990, the population of Chula Vista was about 1.6 times the 1980 population. The population in 1990 was 135,000. What was the approximate population in 1980?

Define p = Chula Vista's 1980 population

Relate   Chula Vista's 1990 population is 1.6 times the 1980 population.

$$135,000 = 1.6 \times p$$

Write $135,000 = 1.6p$

$$\frac{135,000}{1.6} = \frac{1.6p}{1.6} \qquad \longleftarrow \textbf{Divide each side by 1.6.}$$

$$84,400 = p$$

The population was approximately 83,900 in 1980.

Practice

 Write an expression for each phrase.

1. the sum of x and 5

2. 15 less than a

3. the product of 8 and k

4. n divided by 2.8

Write an equation to model. Then solve each problem.

5. The Sears Tower in Chicago, Illinois has a height of 1454 ft. This is 408 ft taller than the Chrysler Building in New York City. What is the height of the Chrysler Building?

6. The metropolitan population of Los Angeles, California in 1990 was about 1.7 times the metropolitan population of Chicago. The population for the Los Angeles area was 11.9 million. What was the population for the Chicago area?

7. Carlo spends $\frac{1}{4}$ of his monthly income on car insurance. His monthly income is $425. What does he spend each month for car insurance?

8. One-third of the problems on the next math test are multiple choice. The test contains 42 problems. How many multiple-choice questions are on the test?

9. The total bill for a video tape is $25.44 including tax. The tax is $1.44. What is the marked price of the tape?

Lesson 3-2 Modeling and Solving Two-Step Equations Part 1
Using Tiles

Example

Use tiles to solve the equation $3x - 1 = 5$.

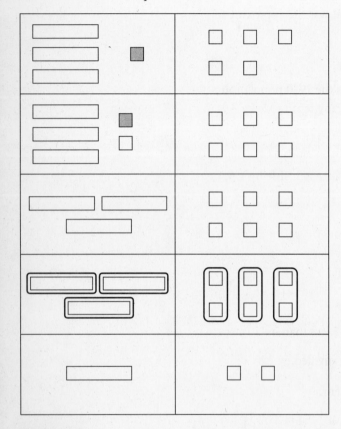

← **Model the equation with tiles.**

$3x - 1 = 5$

← **Add 1 to each side.**

$3x - 1 + 1 = 5 + 1$

← **Simplify by removing zero pairs.**

$3x = 6$

← **Divide each side into three equal groups.**

$$\frac{3x}{3} = \frac{3x}{3}$$

← **Solve for 1x.**

$1x = 2$
$x = 2$

Practice

Evaluate each expression.

1. $2n + 8$ for $n = -5$ **2.** $-5p - 1.5$ for $p = 7$

3. $5 - 2a$ for $a = -3$ **4.** $5x - \frac{1}{3}$ for $x = \frac{2}{3}$

Use tiles to solve each equation.

5. $2x + 3 = 5$ **6.** $4b - 1 = 7$ **7.** $2 + 3a = 11$

8. $2y - 1 = 7$ **9.** $-3 = 3p - 6$ **10.** $3 = 2n - 5$

11. $3r + 1 = -2$ **12.** $5x + 2 = -3$ **13.** $-7 = 5 + 2b$

14. $4t + 5 = 9$ **15.** $0 = 2z - 8$ **16.** $6m + 4 = -8$

17. $3a - 1 = 7$ **18.** $2y - 18 = 44$ **19.** $3x - 1 = 8$

© Prentice-Hall, Inc.

Lesson 3-2 Modeling and Solving Two-Step Equations

Part 2

Using Properties

Example

Solve the equation $6 = -\frac{x}{5} + 10$.

$$6 = -\frac{x}{5} + 10$$

$$6 - 10 = -\frac{x}{5} + 10 - 10 \quad \longleftarrow \text{ Subtract 10 from each side.}$$

$$-4 = -\frac{x}{5}$$

$$-5(-4) = -5\left(-\frac{x}{5}\right) \quad \longleftarrow \text{ Multiply each side by } -5.$$

$$20 = x$$

The solution is 20.

Practice

Solve each equation.

1. $6 + x = 8$ **2.** $2y = 12$ **3.** $12 = -2a$

4. $-\frac{n}{5} = 7$ **5.** $8 = \frac{p}{2}$ **6.** $-3a = -5$

Solve each equation. Check your solutions.

7. $4x + 4 = 24$ **8.** $12 = 5n + 2$

9. $-8 + 3z = -2$ **10.** $7 = 7 - \frac{a}{5}$

11. $-10 - x = 7$ **12.** $\frac{t}{3} - 7.8 = 5$

13. $-4 - c = 15$ **14.** $4x + 3.8 = 7.4$

15. $1 = 6 - 0.25k$ **16.** $1.5x - 6 = 9$

17. $5 - 6a = -1$ **18.** $\frac{m}{8} - 2 = 5$

Use an equation to model and solve each problem.

19. Nicole bought six concert tickets for a total of $114. This included a service charge of $6. How much did each ticket cost?

20. You bought jeans that were on sale for $20 each. The total tax on your purchase was $4. How many jeans did you buy if your total was $84?

21. Jan bought a number of tapes for $7 each. Then she spent $3 more for a poster. If she spent a total of $24, how many tapes did she buy.

22. Kendra works part time at a hospital and is paid by the hour. How much does she earn per hour if her pay was $93 and she worked 20 h?

Lesson 3-3 Combining Like Terms to Solve Equations

Part 1

Combining Like Terms

Terms are **like terms** if they have exactly the same variable factors.
You can combine like terms by adding coefficients.

Example

Simplify the expression $3x - 2 + y - x - 6 - 8y + 5$.

terms with x	terms with y	constant terms	
$3x - 1x$	$1y - 8y$	$-2 - 6 + 5$	← Group like terms.
$2x$	$-7y$	-3	← Then add or subtract.

The simplified expression is $2x - 7y - 3$.

Practice

Simplify.

1. $-8 + (-5)$ **2.** $5 - 12$ **3.** $-6 + 15$

4. $7 - (-3)$ **5.** $-3 - (-7)$ **6.** $-8 - (-8)$

Combine like terms to simplify each expression.

7. $7c - c - c$ **8.** $3m - 7m - m$

9. $9 - y - 7 - 3y$ **10.** $5a - a + 4a$

11. $-7 + 4x + 3 - x$ **12.** $-2b - 7 + b$

13. $k - 4 - 2k + 8$ **14.** $6 - z - 4z + 5$

15. $-4x - y + 4x + y$ **16.** $-3k - 2m + 2k + m$

17. $8a - 10 - 7a + 4$ **18.** $-3r + 5t - r - 4t$

19. $-7e + 8f + 2e - 9f$ **20.** $-6x - 10y + 8z + 4y - 6$

21. $-a + 5.2b + 3a + 3.8b - 10b$ **22.** $7c - 4d - 6.9c + 3d - c$

23. $3x + 5x$ **24.** $-y + 2y + 3 + 2x$

25. $5x + 2x + 6x^2 + x^2$ **26.** $1 + 6m - 2m + 12$

27. $6s - 3s + 6 - 2$ **28.** $7y - 6 - x + 2 - 3x$

29. $8 + 3x^2 + 7 + 2x^2$ **30.** $6m - 19 + 27 - 11n + 4m - 11n$

31. $9b^2 + 20b - 8b^2$ **32.** $-2 + 5g - 4h + 3h + 4 + 7g$

© Prentice-Hall, Inc.

Lesson 3-3 Combining Like Terms to Solve Equations

Part **2**

Solving Equations

Example

You order 5 plain bagels and 8 onion bagels. Each bagel is the same price. With a loaf of bread, which costs $1.50, the total bill is $8.00. Solve the equation $5b + 8b + 1.50 = 8.00$ to find the cost of one bagel.

$$5b + 8b + 1.50 = 8.00$$

$$13b + 1.50 = 8.00 \quad \longleftarrow \text{ Combine like terms.}$$

$$13b + 1.50 - 1.50 = 8.00 - 1.50 \quad \longleftarrow \text{ Subtract 1.50 from each side.}$$

$$13b = 6.50$$

$$\frac{13b}{13} = \frac{6.50}{13} \quad \longleftarrow \text{ Divide each side by 13.}$$

$$b = 0.50$$

Each bagel costs $.50.

Practice

Simplify.

1. $-8a + (-5a)$ **2.** $5x - 12x$ **3.** $-b + 15b$

4. $7y - (-y)$ **5.** $-3a - (-7a)$ **6.** $-8n - (-8n)$

Solve and check each equation.

7. $n + 4n - 11 = 19$ **8.** $9 - y + 6y = -6$ **9.** $60 - 12b + 12 = 0$

10. $g + 8g - 5 = -1$ **11.** $6x + 7.2 - x = 2.4$ **12.** $2c + 7.5c = 57$

13. $a + a + 5 + a + 3 = -1.24$ **14.** $x + 2x + 3x - 7 = -25$

Use an equation to solve each problem.

15. A printer wants to center a 6-in. wide column of text on a page that is 8.5 in. wide. If the margins on the two sides are even, how wide should each margin be?

16 Tapes and discs are on sale for the same price. You buy 4 tapes and 2 discs. You also buy a video that costs $16. The total bill is $82. How much does each tape cost?

17. The ages of four cousins are consecutive integers. Let the integers equal $n, n + 1, n + 2$, and $n + 3$. The sum of their ages is 26. How old is each of the cousins?

18. The perimeters of a square and an equilateral triangle add up to 77 cm. Both figures have sides of the same length. How long is each side?

Lesson 3-4 Using the Distributive Property

Simplifying Variable Expressions

Distributive Property: for all real numbers, *a*, *b*, and *c*,

$$a(b + c) = ab + ac \qquad (b + c)a = ba + ca$$

$$a(b - c) = ab - ac \qquad (b - c)a = ba - ca$$

Examples

$$-6(2x + 7) = -6(2x) - 6(7) \qquad\qquad (6y - 8)(3) = (6y)(3) - 8(3)$$

$$= -12x - 42 \qquad\qquad\qquad\qquad = 18y - 24$$

Practice

Simplify.

1. $-3(-8)$ **2.** $5(-6)$ **3.** $-11(5)$ **4.** $-9(-6)$

5. $-11(20)$ **6.** $-13(-13)$ **7.** $6\left(-\frac{1}{2}\right)$ **8.** $-0.4(20)$

Simplify each expression.

9. $8(x - 3)$ **10.** $3(a - 7)$ **11.** $-8(n + 5)$

12. $-(6b + 2)$ **13.** $-(3t - 7)$ **14.** $-(p + 8)$

15. $(4r - 12)6$ **16.** $(x + 7)(3)$ **17.** $(6 - 5k)(-2)$

18. $(4w - 12)\frac{3}{4}$ **19.** $(g + 8)\frac{1}{2}$ **20.** $\frac{2}{3}(18d - 30)$

21. $16\left(n - \frac{1}{2}\right)$ **22.** $\frac{3}{5}(15x - 5)$ **23.** $(6t - 7)8$

24. $-(4p - 5)$ **25.** $-3(4a + 1.5)$ **26.** $(7 - 3y)(-5)$

27. $-3(a + 2)$ **28.** $(5y + y)(-2)$ **29.** $2(3x - 11)$

30. $7(6 + 2y)$ **31.** $-\frac{3}{4}\left(\frac{2}{3}x - 8\right)$ **32.** $(-5j + 20)5$

33. $(4x - 2)3$ **34.** $(a + 3)4$ **35.** $-2(3y + 2)$

36. $(3a + 4)\frac{1}{2}$ **37.** $-(3 - m)$ **38.** $(4n + 6)2$

39. $(8t - 2)3$ **40.** $-(6a + 3)$ **41.** $\frac{2}{5}(5d - 2)$

42. $(7a - 3)-2$ **43.** $-3(n + 2)$ **44.** $0.5(6 + x)$

45. $(5x + 1)5$ **46.** $4(3 - t)$ **47.** $(8b - 2)7$

48. $-(-4 - b)$ **49.** $(3w + 1)6$ **50.** $\frac{1}{4}(8 - z)$

© Prentice-Hall, Inc.

Lesson 3-4 Using the Distributive Property

Part 2

Solving and Modeling Equations

Example

You have purchased a total of ten compact discs and cassette tapes. The bill is $111. The discs cost $12 each and the tapes cost $9 each. How many of each type did you purchase?

n = number of $12 discs

$10 - n$ = number of $9 tapes

total cost of discs   plus   total cost of tapes   equals   total bill

     $12n$ + $9(10 - n)$ = 111

$$12n + 9(10 - n) = 111$$

$12n + 90 - 9n = 111$ ⟵ **Use the distributive property.**

$3n + 90 = 111$ ⟵ **Combine like terms.**

$3n + 90 - 90 = 111 - 90$ ⟵ **Subtract 90 from each side.**

$3n = 21$

$\dfrac{3n}{3} = \dfrac{21}{3}$ ⟵ **Divide each side by 3.**

$n = 7$

You purchased 7 discs.

$10 - n = 10 - 7 = 3$; you purchased 3 tapes.

Practice

 Simplify.

1. $x + 5(x - 1)$ **2.** $8a - (2a - 3)$

3. $6(y + 4) - 2y$ **4.** $-3(2r - 1) + r$

Solve and check each equation.

5. $2(a + 7) = 16$ **6.** $-5(b + 2) = 30$

7. $2(c - 3) - c = 9$ **8.** $2(d + 3) + d = 12$

9. $-2(3 - 2g) + 4g = 10$ **10.** $23 = 12 - (6 + k)$

Use an equation to model and solve each problem.

11. A table top is rectangular. The table's length is 60 cm more than its width. The perimeter of the table is 240 cm. Find the length and width of the table.

12. Bill bought 10 lb of peanuts and cashews for his party. The cashews cost $7/lb and the peanuts $3/lb. Bill spent a total of $50. How many pounds of each did he buy?

Lesson 3-5 Rational Numbers and Equations

Part 1

Multiplying by a Reciprocal

Example

Kelly's dog weighs 30 lb. This is $\frac{3}{4}$ the weight of Nell's dog. What is the weight of Nell's dog?

Define $w =$ the weight in pounds of Nell's dog

Relate three-fourths weight of Nell's dog equals weight of Kelly's dog

$$\frac{3}{4}w \qquad = \qquad 30$$

Write $\quad \frac{3}{4}w = 30$

$$\frac{4}{3}\left(\frac{3}{4}w\right) = \frac{4}{3}(30) \qquad \longleftarrow \text{ Multiply each side by } \frac{4}{3}, \text{ the reciprocal of } \frac{3}{4}.$$

$$w = 40$$

The weight of Nell's dog is 40 lb.

Practice

 Simplify.

1. $\frac{5}{2}(20)$
2. $\frac{4}{3}(15)$
3. $\frac{3}{2}(30)$

4. $\frac{8}{5}(100)$
5. $\frac{5}{3}\left(\frac{3}{5}x\right)$
6. $-\frac{6}{5}\left(-\frac{5}{6}a\right)$

Solve each equation. Check your answers.

7. $\frac{2}{3}x = 20$
8. $\frac{4}{5}a = 40$
9. $\frac{1}{4}w = 10$

10. $\frac{5}{8}c = 25$
11. $\frac{1}{2}g = 32$
12. $\frac{2}{5}y = 24$

13. $\frac{2}{9}k = 10$
14. $-\frac{2}{5}t = 12$
15. $\frac{3}{10}z = 15$

16. $\frac{1}{3}a = 12$
17. $\frac{4}{5}z = 32$
18. $-\frac{2}{7}b = 12$

Use an equation to solve each problem.

19. The price of $\frac{2}{3}$ lb of bananas is $.30. What is the price of 1 lb of bananas?

20. You ran $\frac{1}{3}$ mi in 2 min. At this rate, how long will it take you to run 1 mi?

21. Corky estimated that it took him 3 h to complete $\frac{3}{4}$ of his homework. How many hours of homework did he have?

22. A football field is 100 yd in length. If a player runs $\frac{1}{4}$ the total length during a play, how many yards did he run?

Lesson 3-5 Rational Numbers and Equations

Part **2**

Multiplying by a Common Denominator

Example

One-half of the class voted for Anita as class representative. One-third of the class voted for Andre. The total number of votes for these two students was 20. How many students are in the class?

Define c = total number of students in the class

Relate one-half of class plus one-third of class is 20 students

$$\tfrac{1}{2}c \qquad + \qquad \tfrac{1}{3}c \qquad = \ 20$$

Write $\quad \tfrac{1}{2}c + \tfrac{1}{3}c = 20$

$$6\left(\tfrac{1}{2}c + \tfrac{1}{3}c\right) = 6(20) \qquad \longleftarrow \text{ \textbf{Multiply each side by 6.}}$$

$$6\left(\tfrac{1}{2}c\right) + 6\left(\tfrac{1}{3}c\right) = 120 \qquad \longleftarrow \text{ \textbf{Use the distributive property.}}$$

$$3c + 2c = 120 \qquad \longleftarrow \text{ \textbf{Simplify each term.}}$$

$$5c = 120 \qquad \longleftarrow \text{ \textbf{Combine like terms.}}$$

$$\tfrac{5c}{5} = \tfrac{120}{5} \qquad \longleftarrow \text{ \textbf{Divide each side by 5.}}$$

$$c = 24$$

There are 24 students in the class.

Practice

Find the lowest common denominator for each pair of fractions.

1. $\tfrac{2}{3}, \tfrac{1}{2}$ 2. $\tfrac{1}{2}, \tfrac{2}{5}$ 3. $\tfrac{1}{4}, \tfrac{2}{3}$

4. $\tfrac{3}{5}, \tfrac{1}{3}$ 5. $\tfrac{3}{4}, \tfrac{5}{8}$ 6. $\tfrac{4}{5}, \tfrac{3}{10}$

Solve and check.

7. $\tfrac{1}{2}a + \tfrac{2}{5}a = 45$ 8. $\tfrac{2}{3}x - \tfrac{1}{2}x = 7$ 9. $\tfrac{n}{5} + \tfrac{n}{2} = 21$

10. $\tfrac{3y}{4} - \tfrac{y}{3} = 40$ 11. $\tfrac{5b - 2}{3} = 11$ 12. $p + \tfrac{2}{3}p = 55$

Use an equation to solve each problem.

13. Chue spends $\tfrac{1}{10}$ of his clarinet practice time on scales. He spends $\tfrac{1}{3}$ of the practice time on sight reading. Yesterday, Chue spent 39 min on scales and sight reading. How many minutes did Chue practice?

14. Sugar accounted for $\tfrac{1}{3}$ of the calories in a cookie. Fat accounted for $\tfrac{1}{2}$ of the calories. The total number of calories from sugar and fat was 150. How many calories were in the cookie?

Lesson 3-6 Using Probability

Part 1

Finding the Probability of Independent Events

Example

Jodi has a compact disc of music from 20 different movies. Twelve tunes are from the 1980s and eight are from the 1990s. Jodi puts his player on random-choice mode. What is the probability that the first tune played will be from the 1980s and the last tune from the 1990s?

$$P(\text{first tune from 1980s}) = \frac{12}{20}$$

$$P(\text{last tune from 1990s}) = \frac{8}{20}$$

$$P(\text{1980s tune and 1990s tune}) = \frac{12}{20} \cdot \frac{8}{20}$$

$$= \frac{6}{25}$$

The probability of the first tune being from the 1980s and last tune from the 1990s is $\frac{6}{25}$.

Practice

Simplify each product.

1. $\frac{3}{10} \cdot \frac{4}{10}$ 2. $\frac{1}{5} \cdot \frac{2}{5}$ 3. $\frac{3}{15} \cdot \frac{4}{15}$

4. $\frac{1}{6} \cdot \frac{5}{6}$ 5. $\frac{3}{8} \cdot \frac{1}{8}$ 6. $\frac{2}{9} \cdot \frac{3}{9}$

A bag contains three red marbles, four blue marbles, and five green marbles. One marble is picked and then placed back in the bag. A second marble is picked. Find each probability.

7. P(red and blue)   8. P(green and blue)   9. P(blue and green)

10. P(red and green)   11. P(red and red)   12. P(red and not red)

Find the probability of each set of independent events.

13. Ten T-shirts are in a drawer. Five are white, two are blue, two are yellow, and one is red. You pick one shirt at random and replace it. After taking a shower you again randomly pick a shirt. What is the probability that you chose a white shirt both times?

14. The letters in *COMMITTEE* are each put on a separate slip of paper and placed in a bag. One letter is drawn and replaced, and then another letter is drawn. What is the probability that an M and then an E are drawn?

15. There are six puppies in a box. Two are black and four are white. One puppy escapes. Scott puts it back. Later he sees that again a puppy has escaped. What is the probability that it was a black puppy the first time and a white puppy the second time?

Lesson 3-6 Using Probability

Part 2

Finding the Probability of Dependent Events

Example

There are eight light bulbs in a carton. Three of them are defective. You choose one light bulb and then another. What is the probability that both bulbs are defective?

The events are dependent.

$P(\text{first bulb defective}) = \frac{3}{8}$ ← **3 of 8 are defective.**

$P(\text{second bulb defective}) = \frac{2}{7}$ ← **2 of the 7 left are defective.**

$P(\text{both are defective}) = \frac{3}{8} \cdot \frac{2}{7}$

$\qquad\qquad\qquad\qquad = \frac{3}{28}$

The probability that both bulbs are defective is $\frac{3}{28}$.

Practice

Simplify each product.

1. $\frac{2}{3} \cdot \frac{1}{2}$ 2. $\frac{2}{5} \cdot \frac{1}{4}$ 3. $\frac{3}{10} \cdot \frac{2}{9}$

4. $\frac{5}{6} \cdot \frac{4}{5}$ 5. $\frac{5}{8} \cdot \frac{4}{7}$ 6. $\frac{4}{9} \cdot \frac{3}{8}$

A desk drawer contains three blue pens, three black pens, and four red pens. You pick one pen at random and give it to a friend. Then you pick another pen. Find each probability.

7. $P(\text{black and red})$ 8. $P(\text{blue and black})$ 9. $P(\text{blue and red})$

10. $P(\text{red and blue})$ 11. $P(\text{red and red})$ 12. $P(\text{blue and not blue})$

Find the probability of each set of dependent events.

13. There are 20 letter tiles left face down on the table. Tim knows that there is one X-tile and one J-tile. Tim picks two tiles. What is the probability that he will pick the X-tile and then the J-tile?

14. Cathy has 12 tapes in her car. Seven tapes are of country-western music, and the other five are of rock music. Without looking, Cathy puts one tape in the player and then chooses another tape to play next. What is the probability that both are tapes of country-western music?

15. Out of all the students in a biology class, exactly two of them earned a grade of 85 on the last test. There are nine girls and six boys in the class. What is the probability that two girls received a grade of 85?

© Prentice-Hall, Inc.

Lesson 3-6 Using Probability

$\bullet$ Part ▼3

Finding Probability Using an Equation

Example

Two students are chosen to represent the class on a dance committee. The probability that the first one chosen is a boy is $\frac{8}{12}$. The probability that the first one chosen is a boy and the second one chosen is a girl is $\frac{8}{33}$. What is the probability that the second student chosen is a girl?

Define x = probability that second student is a girl

Relate P(boy and girl) equals P(boy) times P(girl)

Write $\qquad \frac{8}{33} \qquad = \qquad \frac{8}{12} \qquad \cdot \qquad x$

$\qquad \frac{8}{33} = \frac{8}{12}x$

$\qquad \frac{12}{8}\left(\frac{8}{33}\right) = \frac{12}{8}\left(\frac{8}{12}x\right) \qquad \longleftarrow$ **Multiply each side by $\frac{12}{8}$.**

$\qquad \frac{4}{11} = x$

Practice

Solve each equation.

1. $\frac{2}{3}x = \frac{1}{2}$ **2.** $\frac{3}{5}a = \frac{1}{3}$ **3.** $\frac{1}{2}y = \frac{1}{5}$

4. $\frac{3}{8}n = \frac{1}{5}$ **5.** $\frac{3}{4}c = \frac{3}{5}$ **6.** $\frac{7}{10}r = \frac{3}{8}$

Find $P(B)$ for each problem.

7. $P(A) = \frac{3}{8}$

$P(A \text{ and } B) = \frac{3}{28}$

$P(B) =$

8. $P(A) = \frac{3}{5}$

$P(A \text{ and } B) = \frac{1}{3}$

$P(B) =$

9. $P(A) = \frac{1}{5}$

$P(A \text{ and } B) = \frac{1}{20}$

$P(B) =$

Use an equation to solve each problem.

10. Two pieces of fruit are taken from a bowl. The probability that the first one is an apple is $\frac{1}{3}$. The probability that the two chosen are first an apple and then a banana is $\frac{1}{11}$. What is the probability that the second piece is a banana?

11. There are two multiple-choice questions on the math test. The probability that the first correct answer is choice B is $\frac{1}{5}$. The probability that both correct answers are choice B is $\frac{1}{20}$. What is the probability that the second correct answer is choice B?

Lesson 3-7 Percent Equations

Solving Percent Equations

To model a percent problem with an equation, express each percent as an equation. Three types of percent problems are modeled and solved below. Symbols to remember:

is ⟶ = (equals)

of ⟶ × (times)

what ⟶ n (variable)

Examples

What is 20% of 45?	15 is 30% of what?	What percent of 60 is 15?
$n = 0.20 \times 45$	$15 = 0.3 \times n$	$n \times 60 = 15$
$n = 9$	$\frac{15}{0.3} = \frac{0.3n}{0.3}$	$\frac{60n}{60} = \frac{15}{60}$
	$n = 50$	$n = 0.25$
		$n = 25\%$

Practice

Write each percent as a decimal.

1. 35% **2.** 5% **3.** 125%

Write each decimal as a percent.

4. 0.34 **5.** 0.7 **6.** 0.03

Write an equation to model each problem.

7. What is 10% of 55? **8.** 12 is 20% of what? **9.** What percent of 40 is 25?

10. 60 is 30% of what? **11.** What is 90% of 400? **12.** What percent of 16 is 12?

13. What is 75% of 68? **14.** What percent of 100 is 38? **15.** 40 is 10% of what?

16. What is 25% of 700? **17.** 1 is what percent of 3? **18.** What percent of 240 is 30?

Use an equation to solve each problem.

19. What is 30% of 50? **20.** 6 is 50% of what? **21.** What percent of 18 is 12?

22. 30 is 20% of what? **23.** What percent of 100 is 45? **24.** What is 80% of 200?

25. What is 75% of 88? **26.** 16 is 40% of what? **27.** What percent of 50 is 32?

28. What is 6% of 248? **29.** 30 is what percent of 80? **30.** What percent of 9 is 12?

Lesson 3-7 Percent Equations

Part 2

Writing Equations to Solve Percent Problems

Example

Mee bought several items at the drugstore. She lost the sales slip, but she knew that the tax was $1.20. The tax rate in her state is 5%. What was the cost of her items before tax was added?

Define c = cost of items before tax was added

Relate $1.20 is 5% of the cost

Write $1.2 = 0.05 \times c$

$1.2 = 0.05c$

$\dfrac{1.2}{0.05} = \dfrac{0.05c}{0.05}$

$24 = c$

The total cost of the items, before tax, was $24.00.

Practice

Write an equation to model each problem.

1. What is 35% of 40?

2. 20 is 25% of what?

3. What percent of 10 is 3?

4. 12 is 75% of what?

5. What percent of 30 is 20?

6. What is 80% of 200?

Use an equation to solve each problem.

7. A poll was taken in a town with a population of 15,000. The poll asked for an opinion of the city council's performance for the past year. The number of people who thought that the council had done well was 9,500. What percent of the population voted favorably?

8. Twenty percent of the students in a class have one or more cats as pets. There are six cat owners in this class. How many students are in this class?

9. Vanise earns $45/wk with a part-time job at the local florist. She decides to put 30% of this amount in a savings account. How much does she put in the savings account each week?

10. Kim conducted a survey of several students in her school. She asked 150 students to name their favorite food. Ninety students named pizza as their favorite food. What percent of the students preferred pizza?

11. Thierry's salary increased from $4.50/h to $4.95/h. His new salary is what percent of his old salary?

12. Twenty-five of the numbers from 1 to 100 are prime. What percent of these 100 integers are prime?

Lesson 3-7 Percent Equations

Part
3

Simple Interest

The formula for **simple interest** is $I = prt$. I is the interest, p is the principal, r is the interest rate per year, and t is the time in years.

Example

After three years, Kim has earned $87.20 in simple interest. The annual interest rate is 3.12%. What was Kim's original investment?

$$I = prt$$ ⟵ **Use the formula $I = prt$.**

$$87.20 = p \cdot 0.0312 \cdot 3$$ ⟵ **Substitute 87.20 for I, 0.0312 for r, and 3 for t.**

$$\frac{87.20}{0.0936} = \frac{0.0936p}{0.0936}$$ ⟵ **Divide both sides by 0.0936.**

$$p = 931.62$$

The investment was $931.62.

Practice

Solve for the indicated variable.

1. $3a = 2b$; b **2.** $4b = c + 3$; c **3.** $xy = 0.5$; y

4. $\frac{x}{3} = y$; x **5.** $2m = 4n$; m **6.** $5gh = 15$; h

Model with an equation. Then answer each question.

7. Lee invested $550 for two years at 3.65% simple interest. How much simple interest will he earn?

8. An investor was told that her investment of $900 will earn $85 interest over three years. What is the annual interest rate?

9. The senior class has earned $125 in interest for a four-year investment at 3.7% simple interest. How much did they invest?

10. Maya invested $250 at 3% simple interest for two years . Her brother invested $200 at 3.75% simple interest for two years. Who will earn more in interest?

11. You invested $700 for two years and earned $50 in interest. What was the annual interest rate?

12. A bank has started a new summer promotion. Customers who have received $100 in interest for a two-year investment at 4% simple interest will be eligible to win a free plane ticket. How much money must a customer have invested to be eligible to win?

Lesson 3-8 Percent of Change

$$\text{percent of change} = \frac{\text{amount of change}}{\text{original amount}}$$

Example

On Monday, eight of the students in health class thought that it was important to limit the amount of fat in their diets. After viewing a video on Tuesday, 20 of the students realized that it was important to limit fat. Find the percent of increase in the students who knew the importance of limiting fat.

$$\text{percent of increase} = \frac{\text{amount of change}}{\text{original amount}}$$

$$= \frac{20 - 8}{8} \qquad \longleftarrow \textbf{Substitute.}$$

$$= \frac{12}{8} \qquad \longleftarrow \textbf{Simplify the numerator.}$$

$$= 1.5 \text{ or } 150\% \qquad \longleftarrow \textbf{Divide.}$$

The percent of increase is 150%.

Practice

Find the amount of change.

1. $12 to $15

2. 16 in. to 10 in.

3. 200 to 150

4. 6.5 to 6.05

5. 6 cm to 6.8 cm

6. $5\frac{1}{2}$ ft to $5\frac{7}{8}$ ft

Find each percent of change. Describe the percent of change as a percent of increase or a percent of decrease.

7. $20 to $25

8. 5 ft to 4 ft

9. 20 mi/h to 35 mi/h

10. 12 cm to 6 cm

11. $2.50 to $7.50

12. 120 lb to 132 lb

13. $96 to $147

14. 97, 232 to 118,000

15. 144 lb to 168 lb

16. 18.5 ft to 22.2 ft

17. $.89/lb to $1.19/lb

18. $48 to $42

19. The price of chicken increased from $3.00/lb to $3.60/lb. Find the percent of increase in the price.

20. Over one year, Antonio's time for running the mile improved from 6 min to 5 min. Find the percent of decrease in his time for the mile.

21. Shirts that normally cost $30 each are on sale for $18 each. Find the percent of decrease in the price of these shirts.

22. The number of students in history class increased from 20 to 30. Find the percent of increase in the number of students in this class.

Lesson 4-1 Using Proportions

Using Properties of Equality

Example

Solve $\frac{x}{5} = \frac{3}{4}$.

$$\frac{x}{5} = \frac{3}{4}$$

$$\frac{x}{5} \cdot 20 = \frac{3}{4} \cdot 20 \qquad \longleftarrow \textbf{Multiply each side by a common denominator such as 20.}$$

$$4x = 15$$

$$\frac{4x}{4} = \frac{15}{4} \qquad \longleftarrow \textbf{Divide each side by 4.}$$

$$x = 3.75$$

The solution is 3.75. The ratios $\frac{3.75}{5}$ and $\frac{3}{4}$ are equal.

Practice

Find a common denominator for each pair of fractions.

1. $\frac{2}{3}, \frac{3}{5}$

2. $\frac{1}{4}, \frac{3}{8}$

3. $\frac{3}{4}, \frac{7}{10}$

4. $\frac{1}{2}, \frac{5}{6}$

5. $\frac{5}{9}, \frac{5}{12}$

6. $\frac{7}{8}, \frac{3}{10}$

Solve.

7. $\frac{a}{10} = \frac{3}{5}$

8. $\frac{b}{12} = \frac{3}{4}$

9. $\frac{c}{20} = \frac{1}{4}$

10. $\frac{2}{3} = \frac{d}{12}$

11. $\frac{3}{10} = \frac{e}{4}$

12. $\frac{1}{2} = \frac{f}{8}$

13. $\frac{g}{15} = \frac{2}{3}$

14. $\frac{9}{10} = \frac{h}{30}$

15. $\frac{i}{4} = \frac{75}{100}$

16. $\frac{j}{100} = \frac{3}{20}$

17. $\frac{k}{16} = \frac{1}{4}$

18. $\frac{m}{48} = \frac{7}{12}$

19. $\frac{2}{3} = \frac{n}{10}$

20. $\frac{p}{6} = \frac{3}{8}$

21. $\frac{q}{18} = \frac{7}{10}$

22. $\frac{18}{x} = 6$

23. $\frac{x}{12} = 5$

24. $\frac{190}{d} = 19$

25. $\frac{9}{a} = \frac{16}{144}$

26. $\frac{23}{y} = \frac{69}{6}$

27. $\frac{13t}{26} = \frac{40}{16}$

28. $\frac{8}{11} = \frac{12}{x}$

29. $\frac{x}{6} = \frac{30}{5}$

30. $\frac{28}{r} = \frac{4}{7}$

31. $\frac{8}{a} = \frac{16}{3}$

32. $\frac{5}{24} = \frac{x}{12}$

33. $\frac{18p}{54} = \frac{12}{9}$

34. $\frac{12c}{28} = \frac{15}{7}$

35. $\frac{6}{15} = \frac{n}{40}$

36. $\frac{3}{93} = \frac{5}{x}$

Lesson 4-1 Using Proportions

Using Cross Products

In a proportion, where $b \neq 0$ and $d \neq 0$: If $\frac{a}{b} = \frac{c}{d}$, then $ad = bc$.

Example 1

$\frac{3}{4} = \frac{9}{12}$, so $(3)(12) = (4)(9)$

Example 2

Solve $\frac{x}{4} = \frac{7}{10}$.

$$x(10) = (4)(7)$$ ← **Use cross products.**

$$\frac{10x}{10} = \frac{28}{10}$$ ← **Divide each side by 10.**

$$x = 2.8$$ ← **Simplify.**

The solution is 2.8.

Practice

Solve.

1. $3x = 24$ **2.** $5y = 35$ **3.** $2a = 11$

4. $10n = 23$ **5.** $8c = 20$ **6.** $10e = 45$

Solve and check.

7. $\frac{z}{12} = \frac{3}{4}$ **8.** $\frac{7}{10} = \frac{g}{30}$ **9.** $\frac{2}{3} = \frac{8}{f}$

10. $\frac{h}{30} = \frac{5}{6}$ **11.** $\frac{b}{8} = \frac{3}{10}$ **12.** $\frac{6}{k} = \frac{9}{15}$

13. $-\frac{1}{2} = \frac{7}{d}$ **14.** $\frac{4}{7} = \frac{a}{28}$ **15.** $-\frac{5}{12} = \frac{3}{e}$

16. $\frac{1.5}{2.5} = \frac{c}{10}$ **17.** $\frac{m}{100} = \frac{3}{25}$ **18.** $-\frac{1}{4} = \frac{25}{p}$

19. $\frac{3}{x} = \frac{9}{15}$ **20.** $\frac{x}{4} = \frac{4}{16}$ **21.** $\frac{18}{10} = \frac{a}{5}$

22. $\frac{b}{9} = \frac{24}{54}$ **23.** $\frac{7}{8} = \frac{x}{40}$ **24.** $\frac{9}{2} = \frac{63}{x}$

25. $\frac{11}{x} = \frac{132}{24}$ **26.** $\frac{m}{13} = \frac{10}{65}$ **27.** $\frac{5}{13} = \frac{n}{65}$

28. $\frac{21}{5} = \frac{c}{2.5}$ **29.** $\frac{29}{7} = \frac{x}{1.75}$ **30.** $\frac{17}{2} = \frac{8.5}{k}$

Lesson 4-1 Using Proportions

Part **3**

Solving Percent Problems Using Proportions

Example

A student found that he spends 40% of his monthly income on car insurance. His car insurance is $60 each month. What is his monthly income?

Define m = monthly income

Relate $60 is 40% of what amount?

Write $\dfrac{60}{m} = \dfrac{40}{100}$ ⟵ part
⟵ whole

$60(100) = 40m$ ⟵ **Use cross products.**

$\dfrac{6000}{40} = \dfrac{40m}{40}$ ⟵ **Divide each side by 40.**

$150 = m$

The monthly income is $150.

Practice

 Solve.

1. $\dfrac{3}{4} = \dfrac{n}{100}$ 2. $\dfrac{36}{100} = \dfrac{n}{25}$ 3. $\dfrac{n}{100} = \dfrac{9}{10}$

4. $\dfrac{6}{n} = \dfrac{25}{100}$ 5. $\dfrac{36}{n} = \dfrac{15}{100}$ 6. $\dfrac{n}{100} = \dfrac{7}{5}$

Solve and check.

7. What is 20% of 60?   8. 20 is 40% of what number?   9. What percent of 50 is 14?

10. 18 is 36% of what number?   11. What is 15% of 30?   12. What percent of 100 is 23?

13. 25 is 10% of what number?   14. What percent of 48 is 12?   15. What is 75% of 48?

16. What is 23% of 200?   17. What percent of 200 is 16?   18. 60 is 200% of what number?

Use a proportion to solve each problem.

19. The tax on a used car is $240. This tax is 8% of the base price of the car. What is the base price of the car?

20. There are 24 programs on 4 local television stations from 7 P.M. to 10 P.M. on Monday night. Nine of these programs are comedy shows. What percent of the programs are comedies?

21. In a mixture of concrete, the ratio of sand to cement is 1 : 4. How many bags of cement are needed to mix with 100 bags of sand?

Lesson 4-2 Equations with Variables on Both Sides

Part 1

Using Tiles to Solve Equations

Example

Use tiles to solve the equation $4x - 5 = x + 4$.

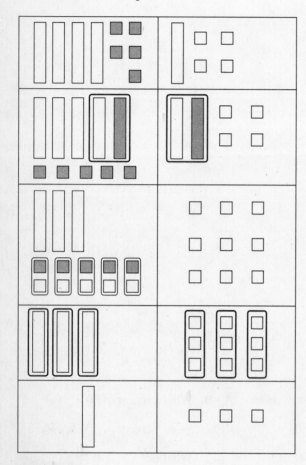

⟵ Model the equation with tiles.

$$4x - 5 = x + 4$$

⟵ Add $-x$ to each side, and simplify by removing zero pairs.

$$4x - 5 - x = x + 4 - x$$

$$3x - 5 = 4$$

⟵ Add 5 to each side, and simplify by removing zero pairs.

$$3x - 5 + 5 = 4 + 5$$

$$3x = 9$$

⟵ Divide each side into three identical groups.

$$\frac{3x}{3} = \frac{9}{3}$$

⟵ Solve for x.

$$x = 3$$

Practice

Solve.

1. $5x - 3x = 14$ 2. $3x + x = 20$

3. $4x - x = 18$ 4. $3x - 5x = 16$

5. $2x + 3x + 7 = 27$ 6. $10x - x - 36 = 0$

Use tiles to solve each equation.

7. $3x + 5 = 2x + 6$ 8. $4x - 3 = 2x + 1$ 9. $x - 2 = 2x - 7$

10. $2x - 5 = x - 1$ 11. $5x + 3 = 4x + 6$ 12. $2x - 3 = 3x - 5$

13. $5x = 2x + 12$ 14. $5a - 9 = 2a$ 15. $2b - 5 = 8b + 1$

© Prentice-Hall, Inc.

Lesson 4-2 Equations with Variables on Both Sides
Part 2

Using Properties of Equality

Example

Solve $7y + 6 = 10y - 3$.

$7y + 6 + 3 = 10y - 3 + 3$	← Add 3 to each side.
$7y + 9 = 10y$	← Simplify each side.
$7y + 9 - 7y = 10y - 7y$	← Subtract $7y$ from each side.
$9 = 3y$	← Combine like terms.
$\frac{9}{3} = \frac{3y}{3}$	← Divide each side by 3.
$3 = y$	← Simplify each side.

Practice

Combine like terms.

1. $3a + 5 - a + 6$ **2.** $6x - 7 + 2x + 7$

3. $3y + y - 27$ **4.** $2n + 8 - 5n - 1$

5. $5t + 7 - t - 2$ **6.** $7q + 11 - 10q + 5$

Solve and check.

7. $8x = 5x + 12$ **8.** $8y = -5y + 65$ **9.** $11a - 36 = 8a$

10. $9t - 18 = 3t$ **11.** $-6r = 10 - 4r$ **12.** $-84 + 15b = 3b$

13. $9p - 5 = 6p + 13$ **14.** $5 - x = x + 9$ **15.** $6 + 10y = 8y + 12$

16. $7n - 9 = 3n + 19$ **17.** $5a - 3 = 7a + 7 + 3a$ **18.** $x + 12 = -2x + 6$

Use an equation to model and solve each problem.

19. Baruch bought a group of 10 tickets. Richard spent $28 more than Baruch to buy 14 tickets. What was the price of 1 ticket?

20. A minivan traveling at 50 mi/h and a motorcycle traveling at 60 mi/h cover the same distance. It takes the minivan 1 h longer to make the trip. How many hours did it take for the motorcycle to make the trip?

21. Sally earns $25,000 per year as a programmer. Her mother's annual income is twice the amount of Sally's. Her brother's total income for one year is $\frac{1}{5}$ of Sally's income. What is their total income?

Lesson 4-2 Equations with Variables on Both Sides

Solving Special Types of Equations

Example 1

An equation has no solution if no value makes the equation true.

Solve $2x + 11 = x - 5 + x$

$2x + 11 = 2x - 5$ ← **Combine like terms.**

$2x + 11 - 2x = 2x - 5 - 2x$ ← **Subtract 2x from both sides.**

$11 = -5$ **Not true for any x!**

Example 2

An equation that is true for every value of the variable is an identity.

Solve $6(5 + 3t) = 18t + 30$

$30 + 18t = 18t + 30$ ← **Use the distributive property.**

$30 + 18t - 18t = 18t + 30 - 18t$ ← **Subtract 18t from each side.**

$30 = 30$ **Always true!**

Practice

Simplify by using the distributive property.

1. $12(x - 5)$ **2.** $6(a + 7)$ **3.** $-2(y + 2)$

4. $-(n - 8)$ **5.** $(p - 5)2$ **6.** $(k + 10)10$

Determine whether each equation has *no solution* or is an *identity*.

7. $2x + 10 = 2x + 6$ **8.** $a + 5 + a = 2a + 5$ **9.** $8t - t + 3 = 7t + 4$

10. $5(y + 7) = 5y + 35$ **11.** $7(q - 3) = 7q - 3$ **12.** $7 - 5x = 11 - 5x$

13. $16 - 4b = 2(8 - 2b)$ **14.** $-3(y + 2) = 6 - 3y$ **15.** $-2(2a - 7) = 14 - 4a$

16. $3d - 15 = 3(5 + d)$ **17.** $5t - 7 = 5t$ **18.** $2x - 3 = x - 3 + x$

Without writing the steps of a solution, tell whether each equation has *one solution*, *no solutions*, or is an *identity*.

19. $6 + 4x = 4x - 5$ **20.** $a + 5 + 2a = 5 + 3a$ **21.** $2x + 3 = x - 5$

Lesson 4-3 Solving Absolute Value Equations

Solving Absolute Value Equations

Example 1

Solve $|x| + 2 = 10$

$|x| + 2 - 2 = 10 - 2$ ← **Subtract 2 from each side.**

$|x| = 8$

$x = 8$ or $x = -8$ ← **The value of x is either 8 or -8.**

Example 2

Solve $|n + 5| = 7$ ← **The value of the expression $n + 5$ is 7 or -7.**

$n + 5 = 7$ or $n + 5 = -7$ ← **Write two equations.**

$n + 5 - 5 = 7 - 5$ or $n + 5 - 5 = -7 - 5$

$n = 2$ or $n = -12$

Practice

Simplify each absolute value.

1. $|-8|$ **2.** $|11 - 5|$ **3.** $|5 - 11|$

4. $|9 + 12|$ **5.** $|-7 + 2|$ **6.** $|-3 - 5|$

Evaluate each expression.

7. $|x + 2|$ for $x = -5$ **8.** $|x - 5|$ for $x = 1$ **9.** $|5 - 3x|$ for $x = 3$

10. $|t + 1|$ for $t = 5$ **11.** $|p - 4|$ for $p = 6$ **12.** $|6 - 2x|$ for $x = 4$

13. $|2k - 9|$ for $k = 1$ **14.** $|m + 3|$ for $m = -4$ **15.** $|r - 1|$ for $r = 6$

Solve and check.

16. $|x| = 5$ **17.** $|a| + 2 = 12$ **18.** $|n| - 6 = 3$

19. $|p| - 9 = -4$ **20.** $15 = |r| + 7$ **21.** $7 = |x| + 7$

22. $|m + 3| = 2$ **23.** $|t - 1| = 6$ **24.** $|x + 8| = 0$

25. $|y| - 5 = -5$ **26.** $|4 - x| = 3$ **27.** $|3s| = 18$

28. $|a + 3| = 2$ **29.** $|5y| = 10$ **30.** $|3k - 6| = 2$

31. $|-3m| = 14$ **32.** $15 = |w + 2|$ **33.** $|t + 1| = 9$

Lesson 4-3 Solving Absolute Value Equations
Part 2

Modeling by Writing Equations

Example

On the last algebra test the average grade was 84. No student's grade varied more than 12 points from the average. What were the maximum and minimum possible grades that students earned?

Define g = possible grade earned by any student in the class

Relate   greatest difference between
average and actual grades   is   12 points

Write $|g - 84|$ =   12

$|g - 84| = 12$ ⟵ **The value of the expression $g - 24$ is 12 or −12.**

$g - 84 = 12$ or $g - 84 = -12$ ⟵ **Write two equations.**

$g = 96$ or $g = 72$

The maximum possible grade was 96 and the minimum possible grade was 72.

Practice

Rewrite each statement as an *or* statement.

1. $|x| = 7$ 2. $|a - 6| = 11$ 3. $|n| = 3$

4. $|r + 2| = 11$ 5. $|t - 1| = 10$ 6. $|2 - e| = 5$

Find the maximum and minimum value of each variable.

7. $|a| = 13$ 8. $|q| = 5$ 9. $|r + 2| = 11$

10. $|x - 7| = 2$ 11. $|y| + 9 = 15$ 12. $|g| - 6 = -3$

13. $|a - 6| = 11$ 14. $10 = 7 + |h|$ 15. $|2 - e| = 5$

Write and solve an equation for each problem.

16. To win the car on a quiz show, you must guess the price within $500. The price of the car is $14,568. What are the maximum and minimum amounts you could guess in order to win the car?

17. This week the temperature has varied from 32°F by as much as 8°F. What are the maximum and minimum possible temperatures for this week?

18. A juice carton claims to contain 64 oz. The manufacturer allows a tolerance of ±2.5 oz per case of 12 cartons. Find the maximum and minimum acceptable amounts of juice per case.

Lesson 4-4 Transforming Formulas

Example

The formula $A = \frac{1}{2}bh$ gives the area of a triangle in terms of the base and the height. Transform the formula to find the height in terms of the area and the base.

$$A = \frac{1}{2}bh$$

$$2A = 2 \cdot \frac{1}{2}bh \qquad \longleftarrow \text{ Multiply each side by 2.}$$

$$2A = bh \qquad \longleftarrow \text{ Simplify.}$$

$$\frac{2A}{b} = \frac{bh}{b} \qquad \longleftarrow \text{ Divide each side by } b, b \neq 0.$$

$$\frac{2A}{b} = h \qquad \longleftarrow \text{ Simplify.}$$

The formula for the height of a triangle is $h = \frac{2A}{b}$.

Practice

 Solve each equation for x.

1. $2x + 5 = 23$ **2.** $5x - 2 = 23$ **3.** $8x = 20$

4. $3(x + 6) = 33$ **5.** $\frac{x}{2} + 5 = -3$ **6.** $\frac{x}{4} = \frac{12}{16}$

Solve each equation for the given variable.

7. $a + b = 16; b$ **8.** $p = a + b + c; a$ **9.** $x - a = 10; x$

10. $6t + z = 12; z$ **11.** $4n + x = 8n; x$ **12.** $\pi = \frac{C}{d}; C$

13. $P = 2l + 2w; w$ **14.** $2x + 5y = 3; x$ **15.** $nx = a; x$

16. $V = lwh; w$ **17.** $\frac{x}{3} = \frac{y}{5}; y$ **18.** $A = P + Prt; r$

Write and solve an equation for each problem.

19. The formula $C = 2\pi r$ gives the circumference C of a circle in terms of the radius r. Remember that π is a constant. Transform the formula to find the radius in terms of the circumference.

20. The formula $F = \frac{9}{5}C + 32$ gives the Fahrenheit temperature F in terms of the Celsius temperature C. Transform the formula to find Celsius temperature in terms of Fahrenheit temperature.

21. The cost to produce a certain number of items is represented by the formula $c = \frac{2}{3}x + 100$. Transform the formula to find the number of items that can be produced in terms of the cost.

Lesson 4-5
Solving Inequalities Using Addition and Subtraction

Part 1

Graphing and Writing Inequalities

Example

The number lines show the solutions of four different inequalities.

$x < 2$

-4 -3 -2 -1 0 1 2 3 4

$x > 2$

-3 -2 -1 0 1 2 3 4 5

$x \leq 2$

-4 -3 -2 -1 0 1 2 3 4

$x \geq 2$

-3 -2 -1 0 1 2 3 4 5

Practice

Label each statement as true or false.

1. $0 > 1$ **2.** $3 > -5$ **3.** $-2 < -6$

4. $-10 < -9$ **5.** $6 \geq 3$ **6.** $8 \leq 8$

Write four numbers that are a solution of each inequality.

7. $y < 1$ **8.** $a > -2$ **9.** $n \leq 0$ **10.** $x < 4$

11. $a \geq 3$ **12.** $b > 0$ **13.** $m \leq -1$ **14.** $t \geq -2$

Rewrite each inequality so that the variable is on the left.

15. $8 \leq x$ **16.** $3 > k$ **17.** $0 \geq t$ **18.** $2 \leq r$

19. $7 > p$ **20.** $6 \geq x$ **21.** $-3 \leq m$ **22.** $8 < d$

Graph each inequality on a number line.

23. $x < 5$ **24.** $d \leq 1$ **25.** $a > -2$ **26.** $y \geq -1$

27. $n \leq 0$ **28.** $6 < b$ **29.** $0 \leq t$ **30.** $w \leq -3$

Lesson 4-5
Solving Inequalities Using Addition and Subtraction

Part 2

Using Addition to Solve Inequalities

Example

Solve $a - 2 > 3$. Graph the solutions on a number line.

$$a - 2 > 3$$

$$a - 2 + 2 > 3 + 2 \qquad \longleftarrow \text{ Add 2 to each side.}$$

$$a > 5 \qquad \longleftarrow \text{ Combine like terms.}$$

The solutions are all numbers greater than 5.

<-+—+—+—+—+—●—+—+—+—+->
0 1 2 3 4 5 6 7 8 9

Check

A number greater than 5 is 7.

$7 - 2 > 3$ is a true statement.

A number less than 5 is 0.

$0 - 2 > 3$ is a false statement.

Practice

Rewrite each set of numbers in order from least to greatest.

1. $-3, 5, -8, 0, 1$ **2.** $-7, -2, -10, -5, -1$ **3.** $6, -6, 0, -2, 2$

4. $-2, 1, -6, 4, 2$ **5.** $-6, -1, -8, 0, -3$ **6.** $5, -5, -2, -6, 1$

Solve each inequality. Graph the solutions on a number line.

7. $y - 7 < 2$ **8.** $x - 5 > -2$ **9.** $a - 8 > 1$

10. $n - 3 \geq 5$ **11.** $k - 1 \leq -3$ **12.** $t - 7 \leq 0$

13. $6 < a - 2$ **14.** $3 > r - 1$ **15.** $-2 \leq m - 3$

16. $x - 0.7 \leq 3.5$ **17.** $b - \frac{1}{2} > 2\frac{1}{2}$ **18.** $0 \geq y - \frac{2}{3}$

19. $a + 4 < 2$ **20.** $c - 1 > 5$ **21.** $z + 2 < -\frac{1}{2}$

22. $x + 2 \leq -3$ **23.** $t - 3 > 0$ **24.** $x - 1 \geq 5$

25. $4 < a - 1$ **26.** $x + 3 > 6$ **27.** $-7 \leq b + 4$

© Prentice-Hall, Inc.

Lesson 4-5
Solving Inequalities Using Addition and Subtraction Part

3

Using Subtraction to Solve Inequalities

Example

If you run 6 mi more this week, you will surpass your goal of 25 mi. How many miles have you already run this week?

Define r = number of miles already run

Relate miles already run plus 6 is greater than 25

Write r + 6 > 25

$r + 6 > 25$

$r + 6 - 6 > 25 - 6$ ← **Subtract 6 from each side.**

$r > 19$ ← **Combine like terms.**

You have already run more than 19 mi this week.

Practice

Use x to write an inequality that represents each expression.

1. more than 12   **2.** less than 16   **3.** 50 or more

4. at least 35   **5.** 75 or less   **6.** at most 100

Solve each inequality. Graph the solutions on a number line.

7. $y + 15 < 20$ **8.** $x + 3 > -1$ **9.** $a + 10 > 4$

10. $n + 3 \geq 5$ **11.** $k + 7 \leq -3$ **12.** $t + 5 \leq 0$

13. $7 < a + 6$ **14.** $9 > r + 11$ **15.** $-2 \leq m + 1$

16. $x - 5 \leq 8$ **17.** $b + \frac{1}{2} > 2\frac{1}{2}$ **18.** $0 \geq y - 3.2$

19. Kim sees a new CD on sale for $12.50. Kim has already bought several discs this month. If she buys this new CD, she will surpass her monthly entertainment budget of $50.00. How much of her entertainment money has Kim already spent?

20. Travis and Neha are loading boxes of clothes onto a truck to be taken downtown and given to a homeless shelter. Each box takes up an area of 4 ft^2. If the bed of the truck has an area of 25 ft^2, how many boxes could they load onto the truck?

© Prentice-Hall, Inc.

Lesson 4-6 Solving Inequalities Using Multiplication and Division

Part 1

Solving Inequalities Using Multiplication

For all real numbers a and b, and for $c > 0$:

$$\text{If } a > b, \text{ then } ac > bc \qquad \text{If } a < b, \text{ then } ac < bc$$

Examples: $3 > -2$, so $3(6) > -2(6)$ $\qquad -4 < -1$, so $-4(2) < -1(2)$

For all real numbers a and b, and for $c < 0$:

$$\text{If } a > b, \text{ then } ac < bc \qquad \text{If } a < b, \text{ then } ac > bc$$

Examples: $3 > -2$, so $3(-6) < -2(-6)$ $\qquad -4 < -1$, so $-4(-2) > -1(-2)$

Example

Solve $-\frac{3}{4}x < 6$. Graph the solution on a number line.

$$-\frac{3}{4}x < 6$$

$$-\frac{4}{3}\left(-\frac{3}{4}x\right) > -\frac{4}{3}(6) \qquad \longleftarrow \text{ Multiply each side by } -\frac{4}{3}. \text{ Reverse the order of the inequality.}$$

$$x > -8 \qquad \longleftarrow \text{ Simplify each side.}$$

The solutions are all numbers greater than -8.

$$\xleftarrow{\;\;\mid\;\;\mid\;\;\mid\;\oplus\;\mid\;\;\mid\;\;\mid\;\;\mid\;\;\mid\;\;\mid\;\;\mid\;\;\mid\;}_{-10\;-9\;-8\;-7\;-6\;-5\;-4\;-3\;-2\;-1\;\;0}\rightarrow$$

Practice

 Multiply each side of each inequality by 5 and write the new inequality.

1. $-3 < 1$ **2.** $8 > 2$ **3.** $-6 < 0$

Multiply each side of each inequality by -5 and write the new inequality.

4. $5 > 4$ **5.** $-3 < -1$ **6.** $3 > 0$

Solve each inequality. Graph the solutions on a number line.

7. $\frac{1}{2}a < 6$ **8.** $\frac{2}{3}n \geq 10$ **9.** $\frac{x}{5} > -2$

10. $-\frac{1}{4}y \leq 5$ **11.** $\frac{t}{-3} < -4$ **12.** $\frac{5}{6}b \geq -10$

13. $12 > \frac{1}{3}x$ **14.** $-11 \leq \frac{n}{7}$ **15.** $-\frac{2}{3}k < 12$

16. $7y < 28$ **17.** $8a < 56$ **18.** $-5z \geq -40$

© Prentice-Hall, Inc.

Lesson 4-6 Solving Inequalities Using Multiplication and Division

Solving Inequalities Using Division

For all real numbers a and b, and for $c > 0$:

If $a > b$, then $\frac{a}{c} > \frac{b}{c}$ If $a < b$, then $\frac{a}{c} < \frac{b}{c}$

Examples: $4 > -4$, so $\frac{4}{2} > \frac{-4}{2}$ $-12 < -9$, so $\frac{-12}{3} < \frac{-9}{3}$

For all real numbers a and b, and for $c < 0$:

If $a > b$, then $\frac{a}{c} < \frac{b}{c}$ If $a < b$, then $\frac{a}{c} > \frac{b}{c}$

Examples: $8 > -2$, so $\frac{8}{-2} < \frac{-2}{-2}$ $-9 < -6$, so $\frac{-9}{-3} > \frac{-6}{-3}$

Example

Suppose you want to save at least $84 for a new pair of running shoes. You decide to save $12 each week. How long will it take you to save at least $84?

Define w = number of weeks

Relate	savings per week	times	number of weeks	is at least	84
Write	12	$\cdot$	w	$\geq$	84

$12w \geq 84$

$\frac{12w}{12} \geq \frac{84}{12}$ ⟵ **Divide each side by 12.**

$w \geq 7$

You will need to save for at least 7 wk.

Practice

Divide each side of each inequality by 2 and write the new inequality.

1. $-4 < 8$ **2.** $8 > 2$ **3.** $-6 < 0$

Divide each side of each inequality by -2 and write the new inequality.

4. $10 > 4$ **5.** $-6 < -4$ **6.** $12 > 0$

Solve each inequality. Graph the solutions on a number line.

7. $3a < 6$ **8.** $5n \geq 10$ **9.** $7x > -14$

10. $-6y \leq 24$ **11.** $-10t < -50$ **12.** $2b \geq -10$

13. Renelle needs to write a research paper of 30 or more pages. She estimates that she can complete 5 pages each day. How many days will it take Renelle to complete the paper?

Lesson 4-7 Solving Multi-Step Inequalities

Solving with Variables on One Side

Part **1**

Example

Solve $-3y - 12 > -78$. Graph the solutions on a number line.

$$-3y - 12 > -78$$

$$-3y - 12 + 12 > -78 + 12 \quad \longleftarrow \text{ Add 12 to each side.}$$

$$-3y > -66 \quad \longleftarrow \text{ Simplify.}$$

$$\frac{-3y}{-3} < \frac{-66}{-3} \quad \longleftarrow \text{ Divide each side by } -3. \text{ Reverse the order of the inequality.}$$

$$y < 22$$

All numbers less than 22 are solutions.

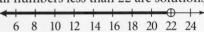

Practice

Use the distributive property to simplify each expression.

1. $2(w + 2)$ **2.** $-6(x + 10)$ **3.** $-5(a - 16)$

4. $(c - 9)5$ **5.** $(t + 8)(-3)$ **6.** $-(r - 12)$

Solve each inequality. Graph the solutions on a number line.

7. $2a + 5 < 13$ **8.** $7 + 6y \geq 55$ **9.** $5x - 9 > 21$

10. $-10g + 15 \leq 95$ **11.** $9 - 3z < 72$ **12.** $3x + 5 + x > -27$

13. $5(w - 10) \geq 105$ **14.** $-2(k + 6) < 38$ **15.** $6t - 1 - 8t \leq 15$

16. $\frac{a}{3} + 7 > 11$ **17.** $\frac{n}{-2} - 12 \leq -20$ **18.** $5g - (g + 1) > 19$

19. Wade wants to buy two shirts and a tie and must spend no more than $60.00. The tie he likes costs $12.50. If the two shirts are the same price, how much can he pay for each shirt?

20. In a new school auditorium, folding chairs 16 in. wide are to be installed side by side to form a row that is no longer than 25 ft. How many chairs will fit in the row?

21. Patio blocks are packaged in bundles containing enough blocks to cover 16 ft². If Sophie wants to build a patio having dimensions 18 ft by 13 ft, how many bundles of blocks should she order?

22. A number k divided by -3 is greater than 1. What is the number?

© Prentice-Hall, Inc.

Name _____ Class _____ Date _____

Lesson 4-7 Solving Multi-Step Inequalities

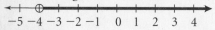

Solving with Variables on Both Sides

Example

Solve $7n + 8 > 2n - 12$. Graph the solutions on a number line.

$$7n + 8 > 2n - 12$$
$$7n + 8 - 2n > 2n - 12 - 2n \quad \longleftarrow \textbf{Subtract } 2n \textbf{ from each side.}$$
$$5n + 8 > -12 \quad \longleftarrow \textbf{Combine like terms.}$$
$$5n + 8 - 8 > -12 - 8 \quad \longleftarrow \textbf{Subtract 8 from each side.}$$
$$5n > -20 \quad \longleftarrow \textbf{Simplify.}$$
$$\frac{5n}{5} > \frac{-20}{5} \quad \longleftarrow \textbf{Divide each side by 5.}$$
$$n > -4$$

All numbers greater than -4 are solutions.

$-5\ -4\ -3\ -2\ -1\ \ 0\ \ 1\ \ 2\ \ 3\ \ 4$

Practice

Combine like terms.

1. $2x + 8 + 5x$ **2.** $3a - 4 - 7a$ **3.** $y + 9y - 14$

4. $n - 23 - 6n$ **5.** $5t + 11 - t$ **6.** $k + 16 + k$

Solve each inequality. Graph the solutions on a number line.

7. $7x - 5 \geq 6x + 6$ **8.** $10a - 7 < 2a + 25$

9. $3b + 4 \leq 2b - 6$ **10.** $2n + 8 > n - 14$

11. $22 - 3r \geq -2r - 11$ **12.** $31 + 14h < 12h + 1$

13. $4 - 3y \geq 5y + 20$ **14.** $2c - 7 \leq 11 - 4c$

15. $11 + 2t < 5t + 26$ **16.** $4 + 3x > 9x - 2$

17. $38 - w \geq 5 - 2w$ **18.** $2g - 23 < 3g - 15$

19. $2y + 5 \geq -y - 4$ **20.** $3m + 6 > -m - 6$

21. $3d - 8 < -6 + d$ **22.** $5a - 14 < -5 + 8a$

23. $4 - 7m \geq m + 4$ **24.** $4 - 9j \leq j + 4$

25. $6n < 4n + 20$ **26.** $11e > 9e + 14$

© Prentice-Hall, Inc.

86 Solving Multi-Step Inequalities Two-Year Algebra Handbook

Lesson 4-8 Compound Inequalities

Part 1

Solving Compound Inequalities Joined by And

Example

Solve $-2 \le x + 3 < 7$. Graph the solutions on a number line.

Write the compound inequality as two inequalities joined by *and*.

$$-2 \le x + 3 \qquad\qquad\qquad \text{and} \qquad\qquad x + 3 < 7$$
$$-2 - 3 \le x + 3 - 3 \qquad\qquad\qquad x + 3 - 3 < 7 - 3 \qquad \longleftarrow \text{ Subtract 3.}$$
$$-5 \le x \qquad\qquad\qquad \text{and} \qquad\qquad x < 4 \qquad \longleftarrow \text{ Simplify.}$$

This can also be written as $-5 \le x < 4$.

The solutions are all the numbers greater than or equal to -5 and less than 4.

Practice

Solve each inequality.

1. $x + 3 < 10$ **2.** $a - 5 > 0$ **3.** $2r \ge 18$

4. $-5t < 30$ **5.** $16 < y + 7$ **6.** $-3 \le -3b$

Rewrite each inequality using the word *and*.

7. $3 < x < 10$ **8.** $-7 \le d - 5 < -2$ **9.** $6 > 2y - 7 \ge -2$

10. $3 < 6 - 3x < 9$ **11.** $8 \le t < 15$ **12.** $5 < 2z + 1 < 7$

Write a compound inequality for each graph.

13.
```
←─┼──┼──⊕──┼──┼──┼──●──┼──┼──→
 -3 -2 -1  0  1  2  3  4  5
```

14.
```
←─┼──⊕──┼──┼──┼──┼──┼──┼──⊕──┼──→
  -1  0  1  2  3  4  5  6  7  8
```

15.
```
←─┼──●──┼──┼──┼──┼──●──┼──┼──┼──→
 -9 -8 -7 -6 -5 -4 -3 -2 -1  0
```

16.
```
←─┼──⊕──┼──┼──┼──●──┼──┼──┼──→
 -4 -3 -2 -1  0  1  2  3  4
```

Solve each inequality and graph the solutions.

17. $-2 < a + 1 < 3$ **18.** $-1 < b - 1 \le 2$ **19.** $-3 < -3 + c < 2$

20. $-1 \le 2 + d < 4$ **21.** $-1 \le 2e + 1 < 5$ **22.** $-3 < 2f - 1 < 9$

23. $-5 \le 3g + 1 < 4$ **24.** $-5 < 3h - 2 \le 7$ **25.** $-6 < -2k < 4$

© Prentice-Hall, Inc.

Lesson 4-8 Compound Inequalities

Solving Compound Inequalities Joined by Or

Example

Solve $2y + 7 < -3$ or $-3y - 4 \leq 5$. Graph the solutions.

$$2y + 7 < -3 \qquad \text{or} \qquad -3y - 4 \leq 5$$

$$2y + 7 - 7 < -3 - 7 \qquad\qquad -3y - 4 + 4 \leq 5 + 4$$

$$2y < -10 \qquad\qquad\qquad -3y \leq 9$$

$$\frac{2y}{2} < \frac{-10}{2} \qquad\qquad\qquad \frac{-3y}{-3} \geq \frac{9}{-3}$$

$$y < -5 \qquad \text{or} \qquad\qquad y \geq -3$$

The solutions are all numbers that are less than -5 or greater than or equal to -3.

```
←—+—+—+—⊕—+—●—+—+—+—→
 -8 -7 -6 -5 -4 -3 -2 -1  0
```

Practice

Rewrite each of the following as an *or* statement.

1. $h \leq -6$ **2.** $g \geq 5$ **3.** $a + 3 \leq -2$

4. $x - 10 \geq 23$ **5.** $5y \geq 20$ **6.** $3z - 8 \leq 16$

Write a compound inequality for each graph.

7.
```
←—+—+—⊕—+—+—+—●—+—+—→
   0  1  2  3  4  5  6  7  8
```

8.
```
←—+—●—+—●—+—+—+—+—+—→
 -7 -6 -5 -4 -3 -2 -1  0  1
```

9.
```
←—+—+—⊕—+—+—+—⊕—+—→
 -2 -1  0  1  2  3  4  5  6
```

Graph each compound inequality.

10. $x < -2$ or $x > 1$ **11.** $c > 0$ or $c \leq -5$ **12.** $y \leq 2$ or $y > 3$

Solve each inequality and graph the solutions.

13. $a - 1 < -2$ or $a - 1 \geq 2$ **14.** $k + 5 \leq -3$ or $k + 5 \geq 3$

15. $3n + 1 \leq -5$ or $3n + 1 > 7$ **16.** $1 + 2b < -5$ or $1 + 2b > 5$

17. $2w - 1 \leq -3$ or $3 \leq 2w - 1$ **18.** $3x + 5 < -13$ or $-5x - 7 \leq 13$

Lesson 4-8 Compound Inequalities

Part 3

Solving Absolute Value Inequalities

$|x| < 5$ can be expressed as $-5 < x$ and $x < 5$.

All solutions are points whose distance from zero is less than 5.

$-6\ -5\ -4\ -3\ -2\ -1\quad 0\quad 1\quad 2\quad 3\quad 4\quad 5\quad 6$

$|x| > 5$ can be expressed as $x > 5$ or $x < -5$.

All solutions are points whose distance from zero is greater than 5.

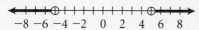

$-8\ -6\ -4\ -2\quad 0\quad 2\quad 4\quad 6\quad 8$

Example

Solve $|x - 2| < 5$. Graph the solutions on a number line.

Write $|x - 2| < 5$ as two inequalities joined by *and*.

$$-5 < x - 2 \qquad \text{and} \qquad x - 2 < 5$$

$$-5 + 2 < x - 2 + 2 \qquad \qquad x - 2 + 2 < 5 + 2$$

$$-3 < x \qquad \text{and} \qquad x < 7$$

The solutions are all numbers greater than -3 and less than 7.

Notice that all solutions are less than 5 units from 2.

$-8\ -6\ -4\ -2\quad 0\quad 2\quad 4\quad 6\quad 8$

Practice

Mini Help

Solve each equation.

1. $|x| = 2$ 2. $|n| = 15$ 3. $|a + 2| = 3$

4. $|g - 6| = 3$ 5. $|4k| = 28$ 6. $|5t| = 0$

Choose a variable and write an absolute value inequality that represents each set of numbers on a number line.

7. all numbers more than 5 units from 0   8. all numbers less than 2 units from 0

9. all numbers more than 3 units from 7   10. all numbers less than 5 units from 8

Express each absolute value inequality as a compound inequality. Solve and graph the solutions on a number line.

11. $|a| > 3$ 12. $|n| \leq 4$ 13. $|b| \geq 7$

14. $|x - 3| < 2$ 15. $|r - 1| \leq 4$ 16. $|w - 7| > 2$

17. $|d + 2| \geq 3$ 18. $|4k| < 28$ 19. $|2c - 6| > 4$

© Prentice-Hall, Inc.

Lesson 4-9 Interpreting Solutions

Solving Inequalities Given a Replacement Set

Example

Solve $-4 \leq 3x < 9$. Then graph the solutions on a separate number line for each replacement set.

    a. the real numbers

    b. the integers

    c. the positive integers

$$-4 \leq 3x < 9$$

$$\tfrac{1}{3}(-4) \leq \tfrac{1}{3}(3x) < \tfrac{1}{3}(9) \qquad \longleftarrow \textbf{Multiply by } \tfrac{1}{3}.$$

$$\tfrac{-4}{3} \leq x < 3 \qquad\qquad \longleftarrow \textbf{Simplify.}$$

a. All real numbers greater than or equal to $\frac{-4}{3}$ and less than 3 satisfy the inequality.

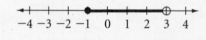

b. The integers that satisfy the inequality are $-1, 0, 1,$ and 2.

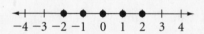

c. The positive integers that satisfy the inequality are 1 and 2.

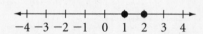

Practice

Which of $-2, -1, 0, 1, 2$ are solutions of each inequality?

1. $x > 0$ **2.** $x \leq 1$ **3.** $x > -5$

4. $|x| = 1$ **5.** $|x| < 2$ **6.** $|x| > 1$

For each inequality, graph the solutions on a separate number line for each replacement set:

    a. the real numbers

    b. the integers

    c. the positive integers

7. $a \leq 5$ **8.** $n > 1$ **9.** $-2 \leq y < 7$

10. $t < -1$ or $t \geq 3$ **11.** $|x| = 2$ **12.** $|x| \leq 2$

13. $|x| > 2$ **14.** $|b - 1| < 4$ **15.** $|3x| \leq 15$

16. $a < 3$ and $a > -1$ **17.** $x \geq -2$ **18.** $|2b| \leq 12$

Lesson 4-9 Interpreting Solutions

Part 2

Determining A Reasonable Answer

Example

The cost of a new jacket that you find attractive is at least $85. You decide that you can save $15 per week for the jacket. How many weeks will it take you to save the money?

Define w = number of weeks

Relate   savings per week   times   number of weeks   is at least   cost of jacket

Write   15   • w ≥   85

$$15w \geq 85$$

$$\frac{15w}{15} \geq \frac{85}{15} \qquad \longleftarrow \textbf{Divide each side by 15.}$$

$$w \geq 5\frac{2}{3} \qquad \longleftarrow \textbf{Simplify.}$$

You must save for at least 6 wk to buy the jacket.

Practice

List each set of integers. If necessary, use … to indicate that a pattern continues.

1. integers between -1.5 and 7
2. positive integers greater than 2.3
3. positive integers less than 5
4. negative integers greater than -6

Write each statement as an inequality.

5. the most that $13x$ can be is 100
6. the least that $5x$ can be is 75
7. $7x$ must be between 10 and 100

Write an inequality and solve each problem.

8. Each apple costs $.29. What is the most number of apples you can buy for $2.00?

9. At the dog show, 8 points are deducted for each command your dog does not execute properly. The lowest grade that will qualify your dog for the next stage of competition is 80 points out of 100. What is the maximum number of commands your dog can miss in order to still qualify for the next round?

10. On the next math test, all students who earn a grade below 80 points out of 100 must do an extra review sheet. Each correct answer on the test is worth 3 points. What is the least number of correct answers you must get to avoid doing the review sheet?

11. A taxi ride in the city starts with a base charge of $2.00. Then for each mile traveled the charge is $.75. How far can you go for $6.00?

Lesson 5-1 Slope

Counting Units to Find Slope

$$\text{slope} = \frac{\text{vertical change (rise)}}{\text{horizontal change (run)}} \quad \text{(reading from left to right)}$$

Example

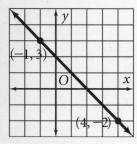

$$\text{slope} = \frac{\text{vertical change}}{\text{horizontal change}} = \frac{-5}{5} \text{ or } -1$$

The slope of the line is -1.

Practice

Simplify each ratio.

1. $\frac{8}{6}$ 2. $\frac{10}{4}$ 3. $\frac{6}{4}$ 4. $\frac{10}{5}$ 5. $\frac{9}{3}$ 6. $\frac{4}{4}$

Find the slope of each line.

7.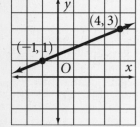

8.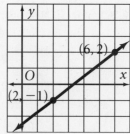

9.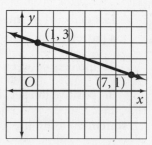

10.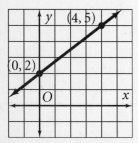

11.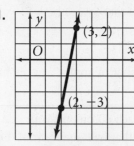

12.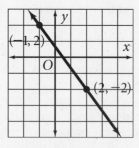

13.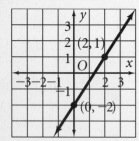

14.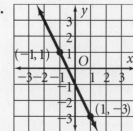

15.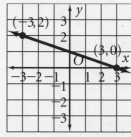

Two-Year Algebra Handbook

Lesson 5-1 Slope

Using Coordinates to Find Slope

slope $= \dfrac{\text{vertical change}}{\text{horizontal change}} = \dfrac{y_2 - y_1}{x_2 - x_1}$, where $x_2 - x_1 \neq 0$

The slope of a *horizontal line* is **0**. The slope of a *vertical line* is **undefined**.

Example

Find the slope of a line through $A(-1, -2)$ and $B(4, 2)$.

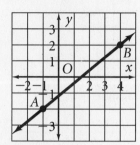

slope $= \dfrac{y_2 - y_1}{x_2 - x_1}$

$= \dfrac{2 - (-2)}{4 - (-1)}$ ← **Substitute $(4, 2)$ for (x_2, y_2) and $(-1, -2)$ for (x_1, y_1).**

$= \dfrac{4}{5}$

The slope of line AB is $\dfrac{4}{5}$.

Practice

Simplify each expression.

1. $-6 - 2$ **2.** $2 - 5$ **3.** $3 - (-1)$

4. $-2 - (-3)$ **5.** $-3 - 5$ **6.** $6 - (-8)$

Find the slope of a line through the two points that are given.

7. $(1, 1), (4, 6)$ **8.** $(0, 2), (4, 0)$ **9.** $(-1, -1), (3, 3)$

10. $(2, -1), (6, 2)$ **11.** $(1, 5), (3, 0)$ **12.** $(-2, 2), (4, 2)$

13. $(1, 0), (1, 3)$ **14.** $(0, 5), (5, 0)$ **15.** $(0, 0), (5, 2)$

16. $(-3, 1), (3, -1)$ **17.** $(-1, -2), (-1, 4)$ **18.** $(1, -2), (7, -2)$

19. $(3, 2), (5, 6)$ **20.** $(3, 8), (1, 4)$ **21.** $(2, 9), (5, 14)$

22. $(4, 7), (7, 11)$ **23.** $(-4, 4), (2, -5)$ **24.** $(-3, 1), (3, -4)$

25. $(-2, 1), (3, -5)$ **26.** $(-5, 2), (2, -4)$ **27.** $(9, -2), (3, 4)$

28. $(6, -3), (1, 2)$ **29.** $(7, -1), (2, 3)$ **30.** $(5, -2), (4, 3)$

31. $(6, 7), (1, 3)$ **32.** $(-1, 3), (2, 4)$ **33.** $(-2, 3), (2, 1)$

34. $(1, 3), (5, -3)$ **35.** $(-5, 2), (3, 2)$ **36.** $(-1, 2), (-1, 4)$

Lesson 5-1 Slope
••• **Part** **3**

Graphing Lines Given a Point and Its Slope

Example

Draw a line through the point $(-1, 4)$ with a slope of $\frac{-2}{5}$.

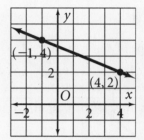

Plot $(-1, 4)$.   →   Find another point . using the slope $\frac{-2}{5}$.   →   Draw a line through the points.

Practice

Mini Help

Determine whether the line through the two given points has a *positive slope*, a *negative slope* or is a *horizontal line* or a *vertical line*.

1. $(2, 3), (4, 6)$ **2.** $(2, 2), (5, 2)$ **3.** $(-1, 0), (3, -1)$

4. $(1, -2), (1, 4)$ **5.** $(-1, 4), (3, 2)$ **6.** $(0, 0), (2, 5)$

Through the given point graph a line with the given slope.

7. $(1, 1)$; slope $= -\frac{2}{3}$ **8.** $(0, 3)$; slope $= \frac{3}{2}$ **9.** $(-2, -1)$; slope $= 2$

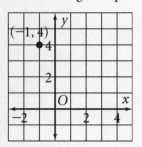   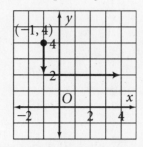   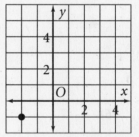

10. $(-2, 0)$; slope $= 0$ **11.** $(2, 2)$; slope $= -1$ **12.** $(3, 5)$; undefined slope

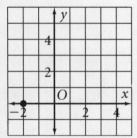   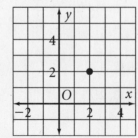   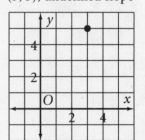

Lesson 5-2 Rates of Change

Finding Rate of Change

You use a rate to find the amount of one quantity per one unit of another. **Rate of change** allows you to see the relationship between two quantities that are changing. The slope of a line is the rate of change between the variables.

Example

Find the rate of change for data graphed on the line. Then explain what the rate of change means in this situation.

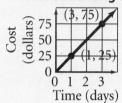

Cost of Renting a Car

Find two points on the graph. Use the points to find the slope, which is also the rate of change.

$$\text{slope} = \frac{\text{vertical change}}{\text{horizontal change}}$$

$$= \frac{75 - 25}{3 - 1}$$

$$= \frac{50}{2}$$

$$= \frac{25}{1}$$

The rate of change is $\frac{25}{1}$, which means that the cost of renting a car increases $25 every day.

Practice

Find each difference.

1. $4 - 7$ **2.** $3 - (-5)$ **3.** $-3 - 8$

4. $-21 - (-14)$ **5.** $-16 - 12$ **6.** $15 - 19$

Find the rate of change for each situation.

7. You drive 120 m in 2 h and 140 m in 3 h.

8. Theater tickets cost $42 for three people and $98 for seven people.

9.   **Cost of Ground Beef**

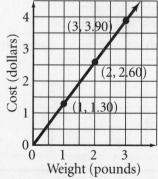

10. Temperature During the Evening

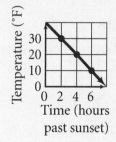

Lesson 5-2 Rates of Change

Using a Table

Part 2

Example

Find the rate of change for buying sandwiches.

Cost of Buying Sandwiches

No. of Sandwiches	Cost
1	$3
2	$6
3	$9
4	$12
5	$15

$$\text{Rate of Change} = \frac{\text{change in dependent variable}}{\text{change in independent variable}}$$

$$= \frac{\text{change in cost}}{\text{change in number of days}}$$

$$= \frac{12 - 6}{4 - 2}$$

$$= \frac{6}{2}$$

$$= \frac{3}{1}$$

The rate of change is $\frac{3}{1}$, which means it costs $3 for each sandwich bought.

Practice

Identify the dependent and independent variables.

1. the length of time a pot of water is on a stove, the temperature of the water

2. the number of miles you can travel, the number of gallons of gasoline purchased

3. the number of calories consumed, the number of candy bars eaten

Find the rate of change.

4. **Cost of Buying Pizzas**

No. of Pizzas	Cost
2	$12
3	$18
4	$24
5	$30
6	$36

5. **Gallons Used per Mile**

Gallons in Tank	Miles Traveled
12	58
11	87
9	145
8	174
7	203

6. **Cost of Buying CDs**

No. of CDs	Cost
2	$26
3	$39
5	$65
6	$78
8	$104

7. **Money Used for Games**

No. of Games Played	Money Remaining
0	$12
4	$11
12	$9
24	$6
32	$4

© Prentice-Hall, Inc.

Lesson 5-2 Rates of Change
Linear Functions

Part **3**

Example

Tell whether the relationship between the data is linear.

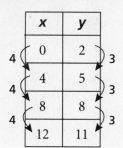

x	y
0	2
4	5
8	8
12	11

rate of change:

(0, 2) to (4, 5) $\frac{3}{4}$

(4, 5) to (8, 8) $\frac{3}{4}$

(8, 8) to (12, 11) $\frac{3}{4}$

4 units ⟶ 3 units

The rate of change is constant between consecutive pairs of data, so the relationship is linear.

Practice

Find the slope for each pair of points.

1. (2, 4) to (3, 3)   **2.** (2, 5) to (5, 7)   **3.** (7, 0) to (10, 1)

4. (4, 6) to (0, 11)   **5.** (1, 6) to (3, 4)   **6.** (9, 7) to (8, 3)

State whether the relationship between the data is linear.

7.

x	y
2	−3
3	−5
5	−9
6	−11

8.

x	y
1	1
3	4
5	6
7	10

9.

x	y
2	14
5	35
9	63
11	77

10.

x	y
−3	15
−1	7
1	−1
2	−5

11.

x	y
13	9
10	6
8	3
5	0

12.

x	y
11	−1
7	3
6	4
4	5

13.

x	y
2	19
5	13
10	3
17	−11

14.

x	y
−1	1
0	0
1	1
2	4

15.

x	y
5	9
4	6
1	−3
−4	−18

16.

x	y
−3	−2
0	0
3	1
9	6

17.

x	y
2	3
4	6
8	12
10	15

18.

x	y
14	17
12	14
8	8
1	−1

Lesson 5-3 Direct Variation

Part 1

Direct Variation

A **direct variation** is a linear function that can be written in the form
$y = kx$, where $k \neq 0$.

↑ constant of variation

Example

Is each equation a direct variation? If it is, find the constant of variation.

a. $3x - 7y = 0$

$\qquad -7y = -3x$ ⟵ **Solve for y.**

$\qquad y = \frac{3}{7}x$

Yes, $y = \frac{3}{7}x$ is in the form $y = kx$. The constant of variation is $\frac{3}{7}$.

b. $3x - 7y = 8$

$\qquad -7y = -3x + 8$ ⟵ **Solve for y.**

$\qquad y = \frac{3}{7}x - \frac{8}{7}$

No, the equation is *not* in the form $y = kx$.

Practice

Solve each equation for y.

1. $0 = x - y$ **2.** $3y = x$ **3.** $y + 7 = 5x$

4. $\frac{1}{2}y = x$ **5.** $4x + y = 9$ **6.** $\frac{5}{3}y = x$

7. $-2x + y = 7$ **8.** $3x - 4y = 9$ **9.** $x = 4y$

Is each equation a direct variation? If it is, find the constant of variation.

10. $2x + 5y = 0$ **11.** $x + 4y = -3$ **12.** $7x - 6y = 0$

13. $-9x + y = -8$ **14.** $-7x - 4y = 0$ **15.** $5x - y = 1$

16. $-6x - y = -8$ **17.** $-2x + 9y = 0$ **18.** $x + y + 1 = 1$

19. $x - 4y = 0$ **20.** $x + y - 6 = 0$ **21.** $y = 6x$

22. $10x - y = 4$ **23.** $x + 5y = 10$ **24.** $x - 8y = 0$

25. $xy = 10$ **26.** $\frac{x}{y} = 7$ **27.** $y = \frac{18}{x}$

Assume y varies directly with x. Find the constant of variation.

28. $x = 2; y = 3$ **29.** $x = -3; y = 6$ **30.** $x = \frac{1}{2}; y = \frac{2}{3}$

31. $x = 5; y = 2$ **32.** $x = 1; y = -4$ **33.** $x = -2; y = -8$

34. $x = 3; y = 9$ **35.** $x = 2; y = 10$ **36.** $x = 4; y = 6$

37. $x = -1; y = -2$ **38.** $x = -3; y = -2$ **39.** $x = 4; y = -3$

Lesson 5-3 Direct Variation

Part 2

Using the Constant of Variation to Write Equations

Example

The area of a circle varies directly with the diameter of the circle. When the diameter is 50 cm, the area is 157 cm^2. Find the constant of variation to write an equation for the relationship between area and diameter.

x = diameter of the circle

y = area of a circle

$y = kx$	← **General form of a direct variation**
$157 = k(50)$	← **Substitute 50 for *x* and 157 for *y*.**
$\frac{157}{50} = \frac{50k}{50}$	← **Solve for *k*.**
$3.14 = k$	← **Constant of variation**
$y = 3.14x$	← **Substitute 3.14 for *k* to write an equation.**

Practice

 Divide.

1. $\frac{42}{7}$ 2. $14 \div 6$ 3. $\frac{7}{20}$

4. $5.25 \div 15$ 5. $48 \div 8$ 6. $\frac{7}{3}$

For each table, tell whether *y* varies directly with *x*. If it does, write an equation for the relationship between the data.

7.
x	y
6	2
15	5
1.8	0.6

8.
x	y
4	10
6	15
3	6

9.
x	y
−8	14
20	−35
2	$-\frac{7}{2}$

10. The distance measured in centimeters varies directly with the distance measured in inches. When the distance is 3 in., the distance is 7.62 cm. Write an equation for the relationship between centimeters and inches.

11. It is a hot summer day and you decide to buy a can of soda. The volume of the can of soda measured in milliliters varies directly with the volume measured in fluid ounces. When the volume is 12 fl oz, the volume is 355 ml. Write an equation for the relationship between milliliters and fluid ounces.

12. The amount of interest earned on a savings account is directly proportional to the amount of money in the account. If $5000 earns $350 interest, how much interest is earned on $8000?

Lesson 5-3 Direct Variation

Part **3**

Using Proportions

Example

Suppose that to lose 1 lb of fat, you need to burn 3500 cal. If you want to lose 10 lb, how many calories must you burn?

Define $pound_1 = 1$ $pound_2 = 10$
         $calories_1 = 3500$ $calories_2 = x$

Write $\dfrac{pound_1}{calories_1} = \dfrac{pound_2}{calories_2}$ ← **Use a proportion.**

         $\dfrac{1}{3500} = \dfrac{10}{x}$ ← **Substitute.**

         $(1)(x) = (10)(3500)$ ← **Cross multiply.**

         $x = 35{,}000$

You need to burn 35,000 cal to lose 10 lb.

Practice

Solve.

1. $\dfrac{x}{56} = \dfrac{1}{7}$ 2. $\dfrac{-1}{5} = \dfrac{-7}{r}$ 3. $3 = \dfrac{p}{27}$

4. $\dfrac{13}{t} = \dfrac{1}{4}$ 5. $\dfrac{-m}{3} = -4$ 6. $\dfrac{8}{b} = \dfrac{1}{6}$

Suppose the ordered pairs in each exercise are for the same direct variation. Find each missing value.

7. $(4, 7)$ and $(12, y)$ 8. $(6, 11)$ and $(x, 16.5)$ 9. $(5, -2)$ and $(x, 10)$

10. $(2, -3)$ and $(-10, y)$ 11. $(x, 6)$ and $(1, 2)$ 12. $(3, y)$ and $\left(\dfrac{9}{7}, 6\right)$

13. $(3, 9)$ and $(x, 15)$ 14. $(2, 7)$ and $(-4, y)$ 15. $(-1, 2)$ and $(x, -5)$

16. A dessert recipe serves 4 people and requires 1 c of sugar. Suppose you have invited 25 people over to your house for a dinner party. How many cups of sugar do you need to make this dessert for the dinner party?

17. The volume in liters of air in a balloon varies directly with temperature in degrees Kelvin. A balloon has a volume of 0.50 L at 298° K. What is the volume at 330° K?

18. An inch on a map represents 23 mi. How far is Houston from Austin if the cities are 7.13 in. apart on the map?

19. An object weighs less on the moon than on earth. A 100-lb object from earth would weigh 16 lb on the moon. How much would a 130-lb astronaut weigh on the moon?

20. A party punch contains 2 qt of pineapple juice, 3 pt of orange juice, $2\frac{1}{2}$ c of cranberry juice, and $1\frac{3}{4}$ c of ice. Find the following ratios. **a.** pineapple juice to ice **b.** cranberry juice to orange juice **c.** orange juice to pineapple juice

Lesson 5-4 Slope-Intercept Form

Part 1

Defining Slope-Intercept Form

The slope-intercept form of a linear equation is $y = mx + b$

↓ slope

↑ y-intercept

Example

Graph $-2x + y = 1$.

$y = 2x + 1$ ← Rewrite the equation in slope-intercept form.

slope ↑   ↑ y-intercept

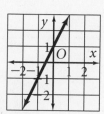

The y-intercept is 1. This means the graph will cross the y-axis at 1. So plot a point at $(0, 1)$. The slope is 2. Remember that slope describes the steepness of the line and is defined as the vertical change over the horizontal change. Rewrite 2 as $\frac{2}{1}$ or $\frac{-2}{-1}$. Use the slope to plot more points. Draw a line through the points.

Practice

Mini Help

For each slope, describe the vertical change and horizontal change in two different ways.

1. 3   **2.** -4 **3.** 0   **4.** $-\frac{1}{2}$ **5.** $\frac{2}{3}$

6. $-\frac{5}{2}$ **7.** 9   **8.** $\frac{4}{7}$ **9.** -2 **10.** 6

Give the slope and y-intercept of each linear equation.

11. $y = 2x - 3$ **12.** $y = -4x + 7$ **13.** $y = -2x - 8$

14. $y = x + 3$ **15.** $y = 9$ **16.** $y = 3x + 2$

17. $y = -x + 5$ **18.** $y = 7x$ **19.** $y = 6x + 5$

20. $y = -9x + 2$ **21.** $y = 5x - 13$ **22.** $y = -\frac{1}{2}x - 7$

23. $y = 14$ **24.** $y = \frac{3}{4}x + 6$ **25.** $y = -11x$

26. $y = -2x$ **27.** $y = -5x$ **28.** $y = 3x + 6$

Rewrite each equation in slope-intercept form. Graph the equation.

29. $3x + y = 2$ **30.** $-4x + y = -1$ **31.** $6x + 2y = -2$

32. $-9x + 3y = 6$ **33.** $2x + 3y = 4$ **34.** $3x - 2y = 8$

35. $-2x + 2y = 4$ **36.** $-4x - 2y = 5$ **37.** $6x - 4y = -9$

38. $-8x + 3y = 7$ **39.** $2x + 6y = -8$ **40.** $-2x - 6y = 12$

Name _____ Class _____ Date _____

Lesson 5-4 Slope-Intercept Form

Writing Equations

Example

Write an equation for the line.

Step 1 Find the y-intercept. The line crosses the y-axis at 1.
The y-intercept is 1 and the point $(0, 1)$ lies on the line.

Step 2 Find another point. The point $(-1, -1)$ also lies on the line.

Step 3 Using these two points, find the slope.
Slope $= \dfrac{-1-1}{-1-0} = \dfrac{-2}{-1} = 2$

Step 4 Using $b = 1$ and $m = 2$, write the equation of the line in slope-intercept form.

$$y = mx + b$$

$$y = 2x + 1$$

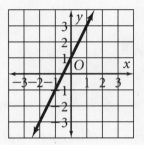

Practice

Substitute 2 for z.

1. $A = \dfrac{bh}{z}$ **2.** $C = z\pi r$ **3.** $P = zl + zw$

4. $y = zx$ **5.** $y = z$ **6.** $y = zx + z$

Write an equation for each line.

7. 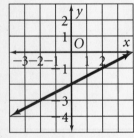   **8.** 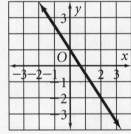   **9.**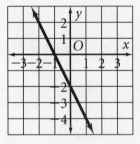

Write an equation of a line for each given slope and y-intercept.

10. $m = \dfrac{1}{4}, b = 2$ **11.** $m = \dfrac{2}{3}, b = -2$ **12.** $m = 0, b = 5$

13. $m = -3, b = 0$ **14.** $m = -\dfrac{1}{2}, b = -3$ **15.** $m = -\dfrac{1}{3}, b = 1$

16. $m = 1, b = 5$ **17.** $m = 6, b = -8$ **18.** $m = \dfrac{2}{3}, b = \dfrac{1}{3}$

19. $m = -3, b = 2$ **20.** $m = -\dfrac{1}{2}, b = -4$ **21.** $m = 5, b = 0$

102 Slope-Intercept Form   Two-Year Algebra Handbook

Lesson 5-5 Writing the Equation of a Line

Example

Write an equation of a line through the points $(3, 4)$ and $(-2, 1)$.

Step 1 Find the slope using the two ordered pairs.

$$\text{slope} = \frac{y_2 - y_1}{x_2 - x_1}$$

$$= \frac{1 - 4}{-2 - 3} \qquad \longleftarrow \textbf{Substitute } (-2, 1) \textbf{ for } (x_2, y_2) \textbf{ and } (3, 4) \textbf{ for } (x_1, y_1).$$

$$= \frac{-3}{-5} = \frac{3}{5}$$

Step 2 Find the y-intercept using the slope and one of the ordered pairs.

$$y = mx + b$$

$$4 = \left(\tfrac{3}{5}\right)(3) + b \qquad \longleftarrow \textbf{Substitute } (3, 4) \textbf{ for } (x, y) \textbf{ and } \tfrac{3}{5} \textbf{ for } m. \textbf{ Solve for } b.$$

$$4 = \tfrac{9}{5} + b$$

$$\tfrac{11}{5} = b \qquad \longleftarrow \textbf{\textit{y}-intercept}$$

Step 3 Substitute values for m and b to write an equation.

$$y = mx + b$$

$$y = \tfrac{3}{5}x + \tfrac{11}{5} \qquad \longleftarrow \textbf{Substitute } \tfrac{3}{5} \textbf{ for } m \textbf{ and } \tfrac{11}{5} \textbf{ for } b.$$

Practice

Simplify.

1. $12 - (-9)$ 2. $-8 - (-4)$ 3. $0 - (7)$

4. $0 - (-11)$ 5. $10 - (6)$ 6. $-3 - (5)$

7. $-9 - 7$ 8. $2 - (-3)$ 9. $7 - 4$

Determine the slope of the line passing through each pair of points.

10. $(7, 5); (-2, 4)$ 11. $(8, 6); (-3, 5)$ 12. $(9, -2); (-3, -4)$

13. $(-7, -2); (-5, -6)$ 14. $(0, -4); (5, 5)$ 15. $(5, 0); (0, 5)$

16. $(-5, -3); (-3, 1)$ 17. $(7, 4); (-2, -6)$ 18. $(1, 0); (0, 0)$

Write an equation of a line through the given points.

19. $(7, 8); (-7, -3)$ 20. $(1, -4); (-2, 8)$ 21. $(-2, 6); (-4, 5)$

22. $(-1, -11); (4, -7)$ 23. $(5, 0); (3, 2)$ 24. $(9, 6); (8, 5)$

25. $(4, -3); (6, 0)$ 26. $(-7, -2); (5, -1)$ 27. $(0, 5); (9, -2)$

28. $(1, 1); (2, 3)$ 29. $(0, 0); (-3, 4)$ 30. $(0, 1); (-1, 1)$

31. $(0, 0); (-1, -2)$ 32. $(3, 4); (-2, 5)$ 33. $(5, -1); (3, -2)$

34. $(-3, -4); (-5, -6)$ 35. $(2, 3); (-1, 5)$ 36. $(3, 1); (6, 2)$

Lesson 5-6 Scatter Plots and Equations of Lines

Part **1**

Trend Line

Example

Find the equation of a trend line for the data.

Step 1 Draw a scatter plot. Use a straight
edge to draw a trend line.

x	y
2	16
4	12
7	14
10	10
13	7
16	6

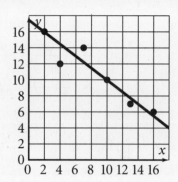

Step 2 Estimate two points on your trend line to find the slope and
write an equation.

slope $= \dfrac{y_2 - y_1}{x_2 - x_1} = \dfrac{10-16}{10-2} = \dfrac{-6}{8} = \dfrac{-3}{4} = -0.75$ ← **Find the slope.**

$y = mx + b$ ← **Use slope-intercept form to find y-intercept.**

$16 = (-0.75)(2) + b$ ← **Substitute −0.75 for m and (2, 16) for (x, y).**

$16 = -1.5 + b$ ← **Solve for b.**

$17.5 = b$ ← **y-intercept**

$y = -0.75x + 17.5$ ← **Substitute −0.75 for m and 17.5 for b.**

Practice

Plot each point.

1. $(1, 3)$ **2.** $(2, -5)$ **3.** $(-4, 7)$

4. $(0, -6)$ **5.** $(-1, -4)$ **6.** $(3, 0)$

Find the equation of a trend line for each set of data.

7.

x	y
−1	3
0	1
1	0
2	1
3	−1

8.

x	y
−5	−8
−3	−7
0	−2
2	−1
4	0

9.

x	y
−2	5
−1	4
0	6
1	5
2	7

10.

x	y
−4	6
−2	4
0	3
2	2
4	0

11.

x	y
−5	−5
−4	−2
−2	2
−1	3
0	5

12.

x	y
−1	−5
1	−3
2	−1
4	1
5	3

Lesson 5-6 Scatter Plots and Equations of Lines

Part 2

Line of Best Fit

Example

Use a calculator to find the equation of the line of best fit for the data below.
Is there a strong correlation between the data?

Product *XYZ* Sales (in millions)

Year	1987	1988	1989	1990	1991	1992	1993	1994	1995
Sales	$1.02	$1.15	$1.30	$1.45	$1.60	$1.75	$1.90	$2.10	$2.30

Step 1 Let 87 represent 1987. Enter the data for years and then the data for sales in your graphing calculator.

Step 2 Find the equation for the line of best fit.

```
LinReg
 y=ax+b
 a=.1578333333        ← slope
 b=-12.74394444       ← y-intercept
 r=.9979965048        ← correlation coefficient
```

The equation for the line of best fit is $y = 0.1578333333x - 12.74394444$.
Since r is very close to 1, there is a strong correlation between the data.

Practice

Graph each equation.

1. $y = x + 2$ **2.** $y = x - 1$ **3.** $y = 2x + 1$

4. $y = -x$ **5.** $y = -x - 2$ **6.** $y = x$

Use a graphing calculator to find the equation of the line of best fit.

7.

x	1	2	3	4	5
y	12	−8	9	5	−3

8.

x	1	2	3	4	5
y	−3	−1	2.5	1.5	3.5

9.

x	1.0	1.5	1.7	2.0	2.3
y	3.8	4.2	5.3	5.8	5.5

10.

x	0	1	2	3	4
y	8	7.5	6.5	5.5	4

11.

x	3.5	3.7	4.0	4.5	5.0
y	9.7	8.6	8.4	7.4	6.8

12.

x	1	2	3	4	5
y	4	−3	6	9	13

13.

x	1	2	3	4	5
y	7	5	−1	8	−5

14.

x	12	15	18	21	24
y	28	50	14	28	36

Lesson 5-7 Ax + By = C Form

Graphing Equations

Part **1**

Example

Graph $5x - 2y = -3$

Step 1 Find the x-intercept by substituting 0 for y and solving for x.

$$5x - 2y = -3$$
$$5x - 2(0) = -3$$
$$5x = -3$$
$$x = -\frac{3}{5}$$

The x-intercept is $-\frac{3}{5}$.

Step 2 Find the y-intercept by substituting 0 for x and solving for y.

$$5x - 2y = -3$$
$$5(0) - 2(y) = -3$$
$$-2y = -3$$
$$y = \frac{3}{2}$$

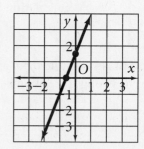

The y-intercept is $\frac{3}{2}$.

Step 3 Plot $\left(-\frac{3}{5}, 0\right)$ and $\left(0, \frac{3}{2}\right)$.

Draw a line through the points.

Practice

Solve for x when $y = 0$.

1. $y = 11x - 22$ **2.** $4x + 7y = -8$ **3.** $y = \frac{1}{3}x + 2$

4. $6x - 2y = 10$ **5.** $y = \frac{3}{2}x - 6$ **6.** $y = -13x + 7$

7. $y = 2x + 3$ **8.** $y = 4x - 3$ **9.** $2y - x = 2$

Graph each equation.

10. $3x + 6y = 12$ **11.** $2x - 4y = -16$ **12.** $-5x + 3y = 15$

13. $-20x - 40y = 10$ **14.** $7x + 9y = -18$ **15.** $-8x + 5y = -16$

16. $2x + y = 3$ **17.** $3x + y = 4$ **18.** $2x + 3y = 7$

19. $x + 2y = 5$ **20.** $2x - y = 3$ **21.** $3x - y = 4$

22. $5x - y = -6$ **23.** $3x - 2y = 4$ **24.** $2x - 3y = -7$

25. $5x - 7y = 12$ **26.** $3x + 4y = -5$ **27.** $4x - 3y = 5$

28. $2x + y = 4$ **29.** $3x + y = 6$ **30.** $2x + 3y = 6$

Lesson 5-7 *Ax* + *By* = C Form

Part
2

Writing Equations

Example

Write the equation of the line with slope $\frac{3}{5}$ through the point $(-2, -4)$.

$$\frac{y - y_1}{x - x_1} = m$$

$$\frac{y - (-4)}{x - (-2)} = \frac{3}{5} \quad \longleftarrow \quad \text{Substitute } (-2, -4) \text{ for } (x, y) \text{ and } \frac{3}{5} \text{ for } m.$$

$$5(y + 4) = 3(x + 2) \quad \longleftarrow \quad \text{Cross multiply.}$$

$$5y + 20 = 3x + 6 \quad \longleftarrow \quad \text{Use the distributive property.}$$

$$-3x + 5y = -14 \quad \longleftarrow \quad \text{Add } -3x \text{ and } -20 \text{ to each side in order to write in } Ax + By = C \text{ form.}$$

Practice

Cross multiply. Then solve for *x*.

1. $\frac{x}{4} = \frac{3}{2}$

2. $-\frac{5}{x} = \frac{7}{9}$

3. $\frac{6}{5} = -\frac{x}{8}$

4. $-\frac{2}{5} = \frac{8}{x}$

5. $\frac{x}{7} = \frac{3}{8}$

6. $\frac{1}{7} = \frac{x}{4}$

7. $\frac{x}{2} = \frac{9}{18}$

8. $\frac{2}{3} = \frac{x}{27}$

9. $\frac{5}{x} = \frac{20}{48}$

Write the equation of the line with the given slope and through the given point.

10. $m = -\frac{3}{7}$; $(5, 4)$

11. $m = 4$; $(-2, -7)$

12. $m = \frac{6}{5}$; $(3, 8)$

13. $m = -2$; $(-9, 6)$

14. $m = \frac{1}{9}$; $(1, -3)$

15. $m = -\frac{8}{3}$; $(-2, -5)$

16. $m = 5$; $(1, 6)$

17. $m = -\frac{3}{4}$; $(1, -2)$

18. $m = -4$; $(-5, 3)$

19. $m = \frac{1}{2}$; $(-3, -9)$

20. $m = -8$; $(7, 0)$

21. $m = \frac{6}{5}$; $(2, 6)$

22. $m = 2$; $(2, 3)$

23. $m = 1$; $(0, 0)$

24. $m = \frac{1}{2}$; $(-1, 1)$

25. $m = 3$; $(0, 4)$

26. $m = 1$; $(-2, 1)$

27. $m = -\frac{2}{3}$; $\left(\frac{1}{2}, 3\right)$

28. $m = -3$; $(4, 5)$

29. $m = -1$; $(0, 0)$

30. $m = -\frac{1}{3}$; $(-3, 4)$

31. $m = -1$; $(1, 2)$

32. $m = \frac{2}{3}$; $(3, 2)$

33. $m = 0$; $(-2, 3)$

34. $m = 1$; $(-1, -1)$

35. $m = -1$; $(2, 2)$

36. $m = -\frac{1}{2}$; $(2, 1)$

37. $m = 3$; $(-3, -2)$

38. $m = \frac{2}{7}$; $(-5, 0)$

39. $m = -\frac{3}{8}$; $(0, 0)$

40. $m = 5$; $(-1, 0)$

41. $m = \frac{7}{2}$; $(-2, -5)$

42. $m = 0$; $(1, -5)$

43. $m = -2$; $(2, 8)$

44. $m = -\frac{8}{3}$; $(0, 0)$

45. $m = 0$; $(-2, 3)$

Lesson 5-8 Parallel and Perpendicular Lines

Part 1

Parallel Lines

Example

Write an equation for a line that contains $(-4, -2)$ and is parallel to the graph of $4x + 3y = 9$.

Step 1 Find the slope of $4x + 3y = 9$.

$$4x + 3y = 9$$
$$3y = -4x + 9$$
$$y = -\frac{4}{3}x + 3 \qquad \longleftarrow \text{ The slope is } -\frac{4}{3}.$$

Step 2 Use the slope intercept form to find the y-intercept.

$$y = mx + b$$
$$-2 = \left(-\frac{4}{3}\right)(-4) + b \qquad \longleftarrow \begin{array}{l}\textbf{Substitute the coordinates } (-4, -2) \textbf{ for } (x, y) \\ \textbf{and } -\frac{4}{3} \textbf{ for } m.\end{array}$$
$$-2 = \frac{16}{3} + b$$
$$-\frac{22}{3} = b \qquad \longleftarrow \text{ The } y\text{-intercept is } -\frac{22}{3}.$$

Step 3 Substitute the values for slope and the y-intercept to write an equation.

$$y = mx + b$$

The equation is $y = -\frac{4}{3}x - \frac{22}{3}$.

Practice

Find the slope of each line.

1. $y = 7x + 4$
2. $x = -1$
3. $-5y + 2x = 3$

4. $3y - 5x = -9$
5. $2y = -8x + 6$
6. $y = 4$

7. $4x - 3y = 12$
8. $-2x + 4y = 8$
9. $-5x + 7y = 2$

Are the lines for each pair parallel?

10. $x - 2y = 5$
$x - 2y = 10$

11. $x + y = 3$
$2x + 2y = 7$

12. $x + y = 4$
$2x + y = 5$

13. $-x + y = 4$
$-2x + y = 2$

14. $2x + 4y = 10$
$4x + 8y = 12$

15. $3x - 6y = 3$
$4x - 7y = 4$

16. $x - y = -6$
$x + y = -2$

17. $2x - y = 7$
$-2x + y = 8$

18. $3x + 4y = 8$
$-9x - 12y = -7$

Write an equation for a line through the given point and parallel to the given line.

19. $(7, 3); y = -x + 2$
20. $(-2, 5); y = 6x - 7$
21. $(-9, -4); y = -5x + 4$

22. $(0, 1); y = 2x - 9$
23. $(-7, 0); y = 3x + 8$
24. $(6, 8); y = -4x - 7$

© Prentice-Hall, Inc.

Lesson 5-8 Parallel and Perpendicular Lines

Part 2

Perpendicular Lines

Example

Write an equation for a line that contains $(3, -5)$ and is perpendicular to the graph of $5x - 4y = -8$.

Step 1 Find the slope of $5x - 4y = -8$.

$$5x - 4y = -8$$
$$-4y = -5x - 8$$
$$y = \frac{5}{4}x + 2 \qquad \longleftarrow \textbf{The slope is } \frac{5}{4}.$$

Step 2 Find the slope of the perpendicular line by taking the opposite reciprocal. The opposite reciprocal of $\frac{5}{4}$ is $-\frac{4}{5}$. The slope of the perpendicular line is $-\frac{4}{5}$.

Step 3 Use the slope-intercept form to find the y-intercept.

$$y = mx + b$$
$$-5 = \left(-\frac{4}{5}\right)(3) + b \quad \longleftarrow \textbf{Substitute the coordinates } (3, -5) \textbf{ for } (x, y) \textbf{ and } -\frac{4}{5} \textbf{ for } m.$$
$$-5 = -\frac{12}{5} + b$$
$$-\frac{13}{5} = b \qquad \longleftarrow \textbf{The } y\textbf{-intercept is } -\frac{13}{5}.$$

Step 4 Substitute the values for slope and the y-intercept to write an equation.

$$y = mx + b$$

The equation is $y = -\frac{4}{5}x - \frac{13}{5}$.

Practice

Find the slope of a line perpendicular to a line with the given slope.

1. $\frac{7}{3}$ 　　2. $-\frac{2}{5}$ 　　3. 4 　　4. -6 　　5. $-\frac{9}{4}$ 　　6. $\frac{5}{8}$ 　　7. 1 　　8. $\frac{4}{3}$ 　　9. $-\frac{3}{8}$

Are the lines for each pair perpendicular?

10. $-4x + y = -5$
$\quad \frac{1}{4}x + y = 8$

11. $2x - 5y = -3$
$\quad 2x + 5y = 4$

12. $3x + 5y = 10$
$\quad 15x + 9y = 18$

13. $x + y = 7$
$\quad -x + y = 3$

14. $x = 5$
$\quad y = \frac{1}{2}$

15. $-3x + 4y = -2$
$\quad 3x + 4y = -12$

Write an equation for a line that contains the given point and is perpendicular to the graph of the given line.

16. $(-12, 8); 12x + 5y = 8$

17. $(-2, 7); -6x + 7y = -3$

18. $(3, 4); x - y = -1$

19. $(-1, -9); -3x - 10y = 7$

20. $(6, -5); 4x - 2y = 9$

21. $(-10, 7); -5x + 4y = 3$

Lesson 5-9 Using the *x*-intercept

Example

Solve $-\frac{4}{3}x + 3 = -5$ by graphing.

Step 1 Write the equation so that one side is 0.

$$-\frac{4}{3}x + 3 = -5$$

$$-\frac{4}{3}x + 8 = 0 \qquad \longleftarrow \textbf{ Add 5 to each side.}$$

Step 2 Replace 0 with *y* to write a function.

$$y = -\frac{4}{3}x + 8$$

Step 3 Graph the function to find the *x*-intercept. The *x*-intercept is 6.

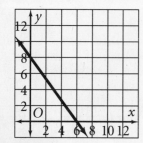

The solution of $-\frac{4}{3}x + 3 = -5$ is 6.

Check $-\frac{4}{3}x + 3 = -5$

$$-\frac{4}{3}(6) + 3 \stackrel{?}{=} -5 \qquad \longleftarrow \textbf{ Substitute 6 for } x.$$

$$-8 + 3 \stackrel{?}{=} -5$$

$$-5 = -5 \ ✔$$

Practice

 Mini Help

Solve for *x*.

1. $2x = 4$

2. $-\frac{8}{5}x + 6 = 1$

3. $\frac{3}{5}x - 7 = 2$

4. $\frac{4}{7}x + 3 = -5$

5. $\frac{1}{3}x = -9$

6. $-3x = 14$

7. $8x - 7 = 41$

8. $2x + 3 = 21$

9. $4x + 27 = 3$

Solve each equation by graphing.

10. $2x - 10 = 12$

11. $\frac{5}{6}x + 2 = 7$

12. $-\frac{15}{2}x - 9 = -11$

13. $\frac{5}{3}x - 4 = -5$

14. $\frac{9}{10}x - 1 = -3$

15. $\frac{4}{7}x + 6 = 8$

16. $3x + 5 = 32$

17. $9x - 1 = 8$

18. $-2x + 5 = 15$

19. $6x + 19 = 7$

20. $5x + 3 = 8$

21. $-6x + 30 = 6$

22. $3x - 5 = 32$

23. $6x + 37 = 17$

24. $\frac{2}{3}x + 5 = 23$

25. $9x - 8 = 10$

26. $\frac{1}{2}x + 1 = 5$

27. $-\frac{5}{4}x + 15 = 23$

28. $8x + 2 = 7$

29. $\frac{7}{10}x + 6 = -1$

30. $\frac{5}{6}x + 3 = 9$

31. $\frac{6}{8}x + 12 = 8$

32. $5x + 3 = 33$

33. $2x + 3 = 7$

34. $2x + 5 = 17$

35. $\frac{3}{4}x - 3 = 18$

36. $-x + 5 = 9$

Lesson 6-1 Solving Systems by Graphing

Solving Systems with One Solution

Example

Solve the system of linear equations by graphing.

$y = 4x - 4$
$y = x + 2$

Graph both equations on the same coordinate grid.

$y = 4x - 4$: slope is 4,

               y-intercept is -4

$y = x + 2$: slope is 1,

               y-intercept is 2

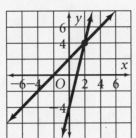

Find the point of intersection.

The lines intersect at $(2, 4)$, so $(2, 4)$ is the solution of the system.

Check See if $(2, 4)$ makes both equations true.

$y = 4x - 4$ $y = x + 2$

$4 \stackrel{?}{=} 4(2) - 4$ ⟵ **Substitute $(2, 4)$ for (x, y)** ⟶ $4 \stackrel{?}{=} 2 + 2$

$4 \stackrel{?}{=} 8 - 4$ $4 = 4$ ✔

$4 = 4$ ✔

It checks, so $(2, 4)$ is the solution of the system of linear equations.

Practice

Determine the slope and y-intercept of each equation.

1. $y = 10x + 4$ **2.** $y = 8x$ **3.** $y = 4.25x + 3.75$

4. $y = 2x - 1$ **5.** $y = -23x - 13$ **6.** $y = \frac{3}{5}x + \frac{1}{2}$

Solve each system of linear equations by graphing. Check your solutions.

7. $y = -x + 4$ **8.** $y = 4x + 9$ **9.** $y = -\frac{1}{2}x - 1$

    $y = x$ $y = \frac{1}{2}x + 2$ $y = -3$

10. $y = \frac{7}{5}x + 2$ **11.** $y = -x + 3$ **12.** $y = -\frac{1}{2}x - 2$

     $y = \frac{2}{5}x - 3$ $y = 2x - 6$ $y = 2x - 2$

13. $y = -2x - 4$ **14.** $y = -x - 3$ **15.** $y = -x + 4$

     $y = x + 5$ $y = 2x - 6$ $y = \frac{1}{4}x - 1$

Lesson 6-1 Solving Systems by Graphing

Part 2

Solving Special Types of Systems

A system of linear equations has *no solution* when the graphs of the equations are parallel. A system of linear equations has *infinitely many solutions* when the graphs are the same line.

Example

Solve the system of linear equations by graphing.

$3y - 5 = 15x + 4$

$\frac{1}{5}y = x - \frac{2}{5}$

First, write each equation in slope-intercept form.

$$3y - 5 = 15x + 4 \qquad\qquad \frac{1}{5}y = x - \frac{2}{5}$$

$$3y = 15x + 9 \qquad\qquad y = 5x - 2$$

$$y = 5x + 3$$

Then, graph each equation on the same coordinate plane.

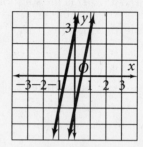

The lines are parallel. There are no points of intersection, so there is no solution.

Practice

Write each equation in slope-intercept form.

1. $\frac{1}{3}y + 2 = x$ **2.** $5y - 2 = 25x - 2$ **3.** $3y = x + 15$

4. $-3 + \frac{1}{9}y = x$ **5.** $\frac{1}{4}y = \frac{3}{4}x + \frac{13}{4}$ **6.** $x = 6y + 5$

Solve each system of linear equations by graphing. Write *no solution* or *infinitely many solutions* where appropriate.

7. $y = 6x - 3$
$\quad\ y = 6x + 5$

8. $3y - x = 6$
$\quad\ y = \frac{1}{3}x + 2$

9. $2y - 4x = 6$
$\quad\ 3y = 6x + 7$

10. $x - 3y = -3y + 1$
$\quad\ \ 6x = 4 + 5x$

11. $x = 28 - 4y$
$\quad\ \ \frac{1}{4}y = -\frac{1}{16}x + \frac{7}{4}$

12. $3y + 34 = 31 + 3x$
$\quad\ \ y = x$

Lesson 6-2 Solving Systems Using Substitution

Part 1

Solving Systems with One Solution

Example

Solve the system
$$y = x - 3$$
$$y = -2x + 3$$
by using substitution.

$$y = x - 3 \quad \longleftarrow \text{ Start with one equation.}$$

$$-2x + 3 = x - 3 \quad \longleftarrow \text{ Substitute } -2x + 3 \text{ for } y \text{ in that equation.}$$

$$-3x = -6 \quad \longleftarrow \text{ Solve for } x.$$

$$x = 2$$

Substitute 2 for x in either equation and solve for y.

$$y = (2) - 3$$

$$y = -1$$

Since $x = 2$ and $y = -1$, the solution is $(2, -1)$.

Check See if $(2, -1)$ satisfies the other equation.

$$-1 \overset{?}{=} -2(2) + 3$$

$$-1 \overset{?}{=} -4 + 3$$

$$-1 = -1 \ \checkmark$$

Practice

Solve each equation.

1. $z - 3 = 5z + 5$ **2.** $3p + \frac{2}{3} = -\frac{7}{3} + 2p$ **3.** $10 - 4b = 6b - 30$

4. $a - 5 = 1 - 3a$ **5.** $\frac{2}{3}d + \frac{4}{3} = \frac{1}{3}d - \frac{14}{3}$ **6.** $2f - 9 = -3f - 5$

Solve each system using substitution. Check your solutions.

7. $y = \frac{4}{3}x$
  $y = 2x - 2$

8. $y = -x + 4$
  $y = \frac{1}{4}x - 1$

9. $y = -2x - 4$
  $y = x + 5$

10. $y = x$
  $y = 2x + 4$

11. $y = -2x - 1$
  $y = x - 4$

12. $y = -\frac{3}{2}x - \frac{1}{2}$
  $y = 3x + 4$

13. $y = -x + 4$
  $y = x$

14. $y = 4x + 5$
  $y = \frac{1}{2}x + 2$

15. $y = -x + 3$
  $y = 2x - 6$

16. $y = 5x + 4$
  $y = 2x - 2$

17. $y = x + 2$
  $y = 4x - 4$

18. $y = 2x - 2$
  $y = -\frac{1}{2}x - 2$

Lesson 6-2 Solving Systems Using Substitution

Part 2

Solving Special Types of Systems

Example 1

Solve the system by using substitution. $x + 4y = 12$
$2x + 8y = 20$

$\quad\quad\quad x = 12 - 4y$ ⟵ **Solve the first equation for x.**

$2(12 - 4y) + 8y = 20$ ⟵ **Substitute $12 - 4y$ for x in the second equation.**

$\quad 24 - 8y + 8y = 20$ ⟵ **Solve for y.**

$\quad\quad\quad\quad 24 = 20$ ⟵ **False!**

Since $24 = 20$ is a false statement, the system has no solution.

Example 2

Solve the system by using substitution. $-27x + 9y = 63$
$21x = 7y - 49$

$\quad 9y = 27x + 63$ ⟵ **Solve the first equation for y.**

$\quad\quad y = 3x + 7$

$21x = 7(3x + 7) - 49$ ⟵ **Substitute $3x + 7$ for y in the second equation.**

$21x = 21x + 49 - 49$ ⟵ **Solve for x.**

$\quad\quad 21 = 21$ ⟵ **Always true!**

Since $21 = 21$ is always a true statement, the system has infinitely
many solutions.

Practice

Solve each equation for the given variable.

1. $5m - n = 7; n$ **2.** $\frac{x}{3} = \frac{y}{6}; x$ **3.** $5y = 2.5x - 25; x$

4. $10a - 5b = 15; b$ **5.** $3g - 0.5f = 4; f$ **6.** $20c + 15d = 10; c$

**Solve each system using substitution. Write *no solution* or *infinitely
many solutions* where appropriate.**

7. $y = 7x + 21$
$\quad -7x + y = 21$

8. $s = 3t + 15$
$\quad 18t = 2 - 6s$

9. $6x - 5.5 = 4.5 - y$
$\quad 1.5x + y = 2.5$

10. $4m - 8n = 3$
$\quad\; m = 5 + 2n$

11. $18x - 3y = 6$
$\quad\; 6x = y + 5$

12. $7.5x = 2.5y - 10$
$\quad\; 4y = 12x + 40$

13. $c = 2b + 4$
$\quad -4b = 8 - 2c$

14. $2v = 2k - 8$
$\quad\; k = v + 5$

15. $y = -\frac{3}{2}x - 1$
$\quad 8y - 12x = -8$

© Prentice-Hall, Inc.

Lesson 6-3 Solving Systems Using Elimination

Part 1

Adding or Subtracting Equations

Example

Solve the system by using elimination.
$$7x + 3y = 25$$
$$2x + 3y = 20$$

First, eliminate one variable.

$$\begin{array}{r} 7x + 3y = 25 \\ \underline{2x + 3y = 20} \\ 5x + 0 = 5 \end{array}$$ ⟵ **Subtract the equations to eliminate *y*.**

$$x = 1$$ ⟵ **Solve for *x*.**

Then, find the value of the eliminated variable.

$$2x + 3y = 20$$ ⟵ **Pick one equation.**

$$2(1) + 3y = 20$$ ⟵ **Substitute 1 for *x*.**

$$2 + 3y = 20$$ ⟵ **Solve for *y*.**

$$3y = 18$$

$$y = 6$$

Since $x = 1$ and $y = 6$, the solution is $(1, 6)$.

Check See if $(1, 6)$ makes the other equation true.

$$7(1) + 3(6) \overset{?}{=} 25$$
$$7 + 18 \overset{?}{=} 25$$
$$25 = 25 ✔$$

Practice

Would you add or subtract the equations to eliminate a variable?

1. $5c + 3d = 23$
$21c - 3d = 13$

2. $3g - 0.5f = 1.25$
$3g - 1.5f = 2.75$

3. $4y - 2.5x = 25$
$-4y - 250x = 15$

4. $13a - b = 15$
$a + b = 2$

5. $\frac{3}{8}t + 8v = 24$
$\frac{3}{8}t + 2v = 6$

6. $2c + 5d = 17$
$4c - 5d = -1$

Solve each system using elimination. Check your solutions.

7. $4c - 3d = 21$
$-4c - 2d = -6$

8. $a + 2b = 3$
$a + b = 2$

9. $6k - 7l = -8$
$3k + 7l = 17$

10. $3w + 2x = 17$
$-3w + 4x = 7$

11. $4g + h = 23$
$-7g - h = -35$

12. $-2x - y = 2$
$4x + y = -12$

13. $m + 3n = 18$
$m + 2n = 17$

14. $10x - 3y = 12$
$-8x + 3y = -6$

15. $4s - 4t = 8$
$4s + t = 3$

Lesson 6-3 Solving Systems Using Elimination

Part 2

Multiplying First

Example

Solve the system by using elimination. $4x + 2y = 8$
$16x - y = 14$

In the first equation, the coefficient of x is 4. In the second equation, the coefficient of x is 16. So multiply the first equation by 4. Then subtract to eliminate x.

$$4x + 2y = 8 \longrightarrow 16x + 8y = 32 \quad \longleftarrow \textbf{Multiply each side of the first equation by 4.}$$
$$\underline{16x - y = 14}$$
$$0 + 9y = 18 \quad \longleftarrow \textbf{Subtract the equations to eliminate } x.$$
$$y = 2 \quad \longleftarrow \textbf{Solve for } y.$$

Then, find the value of the eliminated variable.

$$4x + 2y = 8 \quad \longleftarrow \textbf{Pick one equation.}$$
$$4x + 2(2) = 8 \quad \longleftarrow \textbf{Substitute 2 for } y.$$
$$4x + 4 = 8 \quad \longleftarrow \textbf{Solve for } x.$$
$$4x = 4$$
$$x = 1$$

Since $x = 1$ and $y = 2$, the solution is $(1, 2)$.

Practice

By what should you multiply each equation in the system to eliminate a variable?

1. $3c + 4d = 12$
$12c + 9d = 43$

2. $2g - 7f = 3$
$3g - 5f = 5$

3. $4y - 5x = 22$
$-7y - 25x = 15$

4. $a - b = 15$
$4a + 7b = 2$

5. $\frac{3}{8}t + 8v = 24$
$3t + 5v = 32$

6. $1.75c + 15d = 17$
$4c - 5d = -1$

Solve each system using elimination.

7. $2c + 3d = 26$
$4c - 2d = 4$

8. $3a + 2b = -1$
$-6a - 3b = 6$

9. $3k + 2l = 30$
$7k + 4l = 80$

10. $3w + 4z = -41$
$4w - 3z = 12$

11. $6g - 4h = 11$
$3g + 5h = 23$

12. $x - 5y = 20$
$5x + 22y = 6$

13. $2m + n = -4$
$-m + n = 5$

14. $-x + y = -4$
$4x + 2y = -2$

15. $e + f = 4$
$-5e + 20f = -20$

16. $2x - 5y = 15$
$7x - 5y = -10$

17. $4x + 4y = 12$
$8x - 4y = 24$

18. $-2x - 4y = 8$
$2x - y = 2$

© Prentice-Hall, Inc.

Lesson 6-4 Writing Systems

Example

A small-town movie theater's costs total $330 per night for the projector, the film prints, and other constant costs. In addition to these costs, $2.50 per moviegoer is spent for personnel and concession supplies. Each customer spends $5.50 for a ticket plus an average of $2.50 on concessions. How many people must attend a movie each night for the theater to break even?

Define $c =$ the number of customers
$m =$ the money for expenses or income

Relate Expenses   Income
$2.50 × customers + $330   ($5.50 + $2.50) × customers

Write $m = 2.5c + 330$ $m = 8c$

Choose a method to solve the system. Use substitution, since it is easy to substitute for m using these equations.

$m = 2.5c + 330$

$m = 8c$

$8c = 2.5c + 330$ ⟵ **Substitute 8c for m in the first equation.**

$5.5c = 330$ ⟵ **Subtract 2.5c from each side.**

$c = 60$ ⟵ **Divide each side by 5.5.**

For the theater to break even, 60 people must attend each night.

Practice

 Which method would you prefer to solve each system, substitution or elimination?

1. $7c + d = 124$
$-3c + 2 = d$

2. $32g - 7f = 34$
$-4g - 7f = -8$

3. $-12 - 5x = y$
$9y - 2x = 1$

Solve each problem using a system of linear equations.

4. The ratio of North American steel roller coasters to wooden roller coasters was about 17 to 10 in 1994. There were about 215 roller coasters in existence at that time. About how many were steel? About how many were wooden?

5. A CD store has daily expenses of $198 for personnel and rent. Each CD costs the store $9.00, and is sold for $14.50. How many CDs must the store sell every day to break even?

6. A zoo spends $1,300 daily on feed, staff, and other constant costs, in addition to tour guide expenses of $2 per visitor. If admission is $15, how many visitors must there be each day for the zoo to break even?

Lesson 6-5 Linear Inequalities

Example

Graph $2y - 4x \leq -2$.

Write the inequality in slope-intercept form.

$$2y - 4x \leq -2$$
$$2y \leq 4x - 2$$
$$y \leq 2x - 1$$

Graph the boundary line $y = 2x - 1$.
Points on the boundary line *do* make the inequality true.
So, use a solid line.

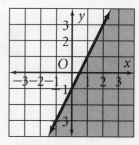

Next, test a point. Use $(0, 0)$.

$$y \leq 2x - 1$$
$$0 \leq 2(0) - 1$$
$$0 \leq -1 \textbf{ False}$$

The inequality is false for $(0, 0)$. So, shade the region not containing $(0, 0)$.

Practice

Determine whether each inequality is true or false for $(0, 0)$.

1. $y < 10x + 4$ **2.** $y \geq 8x$ **3.** $y \leq 4.25x - 3.75$

4. $y > 2x - 1$ **5.** $5 + x - y \leq -3x - 13$ **6.** $y < \frac{3}{5}x + \frac{1}{2}$

Graph each inequality.

7. $y < -5x + 5$ **8.** $y \geq 2x - 5$ **9.** $y > -\frac{1}{2}x - 4$

10. $7 \leq \frac{7}{5}x - y$ **11.** $-3y < -9x + 3$ **12.** $x \leq -\frac{1}{2}y - 2$

Solve each problem by graphing.

13. Suppose you are shopping for wild bird food. Black oil sunflower seeds cost $2/lb, and thistle seeds cost $3/lb. You can spend no more than $15. How many pounds of each can you buy?

14. Lloyd wishes to buy bedding plants for his garden. Small impatiens cost $1, while large ones cost $2. Lloyd has $20 to spend. How many plants can he buy?

Lesson 6-6 Systems of Linear Inequalities

Example

A chef wants to create a new pasta sauce including both garlic and oregano. Each jar of the sauce should include more than 4 g of garlic, but each jar should include no more than 10 g total of the two ingredients. What are the possible amounts of the two ingredients?

Define x = amount of oregano used
y = amount of garlic used

Relate	Garlic used	is more than	4 g.

Write $\quad y \quad\quad > \quad\quad 4$

Use slope-intercept form to graph.

$$y > 4$$
$$m = 0$$
$$b = 4$$

Next, test a point. Use (0, 0).

$$y > 4$$
$$(0) > 4 \text{ \textbf{False.}}$$

Shade above the line.

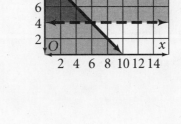

The solutions are all the points in the shaded region above $y > 4$ but below $x + y \leq 10$.

	Total of both ingredients	is no more than	10 g.

$$x + y \quad\quad \leq \quad\quad 10$$

Use intercepts to graph.

$$x + y \leq 10$$
$$(10, 0)$$
$$(0, 10)$$

Test (0, 0).

$$x + y \leq 10$$
$$(0) + (0) \leq 10 \text{ \textbf{True.}}$$

Shade below the line.

Practice

Solve each system of equations.

1. $y = 2x + 4$
$\quad y = -x - 5$

2. $y = 4x$
$\quad 3y = 2x + 20$

3. $y = x - 3$
$\quad -2y = -5x + 3$

Solve each system of linear inequalities by graphing.

4. $y < -2x + 3$
$\quad y \geq 3$

5. $y \geq 5x - 2$
$\quad y \leq 7$

6. $-x - 2y > 2$
$\quad x + y > 2$

Solve by graphing.

7. A gardener wishes to put a new rectangular garden in his backyard. He includes a melon vine, which requires that the garden be at least 4 ft wide. He has only 20 ft of edging, so that is the maximum distance around the garden. What are the possible dimensions of the garden?

Lesson 6-7 Concepts of Linear Programming

Example

Use linear programming. Find the values of x and y that maximize the equation $B = 3x + y$ with the following restrictions:

$$x + y \leq 60$$

$$x \geq 10$$

$$x \leq 15$$

$$y \geq 40$$

STEP 1

Graph the restrictions.

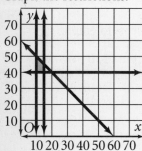

STEP 2

Find coordinates of each vertex.

Vertex

E (10, 40)

F (15, 40)

G (15, 45)

H (10, 50)

STEP 3

Evaluate B at each vertex.

$B = 3x + y$

$B = 3(10) + 40 = 70$

$B = 3(15) + 40 = 85$

$B = 3(15) + 45 = 90$

$B = 3(10) + 50 = 80$

The equation is maximized when $x = 15$ and $y = 45$.

Practice

 Mini Help

Determine which point produces the maximum value of the given equation.

1. $B = 4x + y$
    (3, 3) (5, 3) (4, 4) (3, 5)

2. $C = 3x + 4y$
    (1, 2) (2, 4) (1, 6) (0, 9)

3. $Q = 2x + \frac{1}{2}y$
    (2, 2) (6, 2) (2, 6) (1, 7)

Evaluate the equation to find minimum and maximum values.

4. $y \leq \frac{1}{2}x + 1$
    $x \leq 8$
    $y \geq 1$
    $x \geq 2$
    $A = 3x + y$

5. $x \geq 0$
    $y \geq 0$
    $y \leq -2x + 8$
    $y \geq -3x + 6$
    $B = 8x + 3y$

6. $x \leq 9$
    $x \geq 6$
    $y \leq \frac{1}{3}x + 3$
    $y \geq -\frac{1}{3}x + 3$
    $C = x + y$

Use linear programming to solve.

7. A plastic recycling plant recycles two types of plastic, type A and type B. Due to limits on the machinery, no more than 5 t of type A and no more than 14 t of type B can be recycled each day. The plant makes a profit of $300/t of type A and $280/t of type B. The plant has a total daily capacity of 15 t. How many tons of each type of plastic should the plant recycle to maximize profit?

Name _____ Class _____ Date _____

Lesson 6-8 Systems with Nonlinear Equations

Example

Solve the system of equations. $y = |x - 2|$
                                    $y = \frac{1}{5}x + 2$

Graph each equation.

$y = \frac{1}{5}x + 2$ $y = |x - 2|$

Use $y = mx + b$.   Make a table of values.

$m = \frac{1}{5}$

$b = 2$

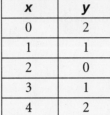

x	y
0	2
1	1
2	0
3	1
4	2

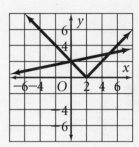

The graphs intersect at $(0, 2)$ and $(5, 3)$.

Check (0, 2)

$y = \frac{1}{5}x + 2$ $y = |x - 2|$

$2 \stackrel{?}{=} \frac{1}{5}(0) + 2$ $2 \stackrel{?}{=} |0 - 2|$

$2 = 2$ ✔ $2 = 2$ ✔

Check (5, 3)

$y = \frac{1}{5}x + 2$ $y = |x - 2|$

$3 \stackrel{?}{=} \frac{1}{5}(5) + 2$ $3 \stackrel{?}{=} |5 - 2|$

$3 = 3$ ✔ $3 = 3$ ✔

Practice

Make a table of values for each equation.

1. $y = x^2 + 2$ **2.** $y = 2x^2 - 3$ **3.** $y = -|x| + 1$

4. $y = 3|x| - 1$ **5.** $y = -2x^2$ **6.** $y = |4x| - 4$

Solve each system by graphing. Check your answers.

7. $y = x + 4$
    $y = 3|x|$

8. $y = x^2 + 3$
    $y = -x + 5$

9. $y = 2x^2 - 5$
    $y = 6x - 5$

10. $y = |4x| - 6$
    $y = 2$

11. $y = -3x^2 + 6$
    $y = -3x$

12. $y = -2|x| + 3$
    $y = 1.2x - 0.2$

13. $y = -x^2 + 9$
    $y = x + 3$

14. $y = \frac{1}{5}x + 3$
    $y = |x - 2| + 1$

15. $y = x + 1$
    $y = 2|x| - 2$

© Prentice-Hall, Inc.

Lesson 7-1 Exploring Quadratic Functions

Quadratic Functions

- For $a \neq 0$, the function $y = ax^2 + bx + c$ is a quadratic function.
- Written in the form $y = ax^2 + bx + c$, it is in standard form.

Example

Write the equation $3x + y = x^2 - 4$ in standard form.
Name the values of a, b, and c.

$$3x + y = x^2 - 4$$

$$y = x^2 - 3x - 4 \longleftarrow \textbf{Subtract } 3x \textbf{ from each side.}$$

$a = 1, b = -3, c = -4$

Practice

Write each equation in standard form.

1. $y = 3x + 2x^2 + 24$ **2.** $y - 2x = 2x^2 - 3$ **3.** $4y = 8x^2 - 12x + 4$

4. $y = 3 + 3x - 3x^2$ **5.** $\frac{1}{3}y = -4x - 3x^2$ **6.** $y = 4 - 8x^2$

Write each equation in standard form and name the values of a, b, and c.

7. $y = 2x^2 + 3 - 18x$

8. $y + 3x = \frac{1}{4}x^2 - 4$

9. $3y = 9x - 3x^2 + 15$

10. $y - 1.75 = 4.5x - 0.75x^2$

11. $2x - 9 = 3x^2 + 2$

12. $5x^2 - 3x = 10 - 6x$

13. $7 - 4x - 3x^2 = 9 - 5x$

14. $-x^2 + 7x = 3x - 9$

15. $x^2 - 5x = 3x + 8$

16. $2x^2 + 3x - 200 = 600 + 3x$

17. $4 = -16x^2 + 6x + 8$

18. $-5x^2 + x + 8 = 3x - 2x^2$

19. $-2x^2 + 6x = 2x - 4 - 3x^2$

20. $4x^2 = -11x - 7$

21. $2x^2 - 6x + 4 = 4$

22. $3x^2 - 8x + 2 = 6$

23. $x^2 - x - 2 = 0$

24. $y^2 - 6y + 9 = 0$

25. $4x^2 = x - 80$

26. $z^2 - 4z = -3$

27. $6m^2 = 72$

28. $x^2 + 8x = 20$

29. $4x^2 + 3x - 1 = 0$

30. $2x^2 + 5x + 3 = 0$

31. $3x^2 - 4x = 2x + 2$

32. $3 + x^2 = 4x$

33. $x^2 + 5x = -6$

34. $2x^2 - 5x - 12 = 0$

35. $4x^2 - 12x = -9$

36. $6x^2 + 7x - 5 = 0$

37. $2x^2 - 5 = 2x$

38. $x^2 - 3x = -2$

39. $3x^2 - 4x = 7$

40. $5x = 3x^2 - 6$

Lesson 7-1 Exploring Quadratic Functions

Part 2

*The role of **a***

Example

Make a table of values and graph $y = 3x^2$, $y = \frac{1}{3}x^2$, and $y = -3x^2$.
Which graph is widest? Narrowest? In which direction does each graph open?

x	$y = 3x^2$	$y = \frac{1}{3}x^2$	$y = -3x^2$
-2	12	$\frac{4}{3}$	-12
-1	3	$\frac{1}{3}$	-3
0	0	0	0
1	3	$\frac{1}{3}$	-3
2	12	$\frac{4}{3}$	-12

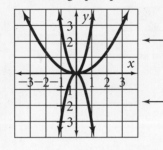

← $a > 0$; these graphs open upward.

← $a < 0$; this graph opens downward.

The smaller the absolute value of a, the wider the graph.
So $y = \frac{1}{3}x^2$ is widest.

$y = 3x^2$ and $y = -3x^2$ are of equal width.

Practice

Put each equation into standard form and find a.

1. $y = 5x + x^2 + 2$
2. $y - 2x = 2x^2 - 3$
3. $3y = 9x^2 - 12x + 3$

4. $y = 3 - x - x^2$
5. $\frac{1}{3}y = -2x^2 + 1 - 4x$
6. $y = 4 - 3x^2$

Order each group of quadratic equations from widest to narrowest graph. Determine in which direction each function opens. Verify your answers by graphing, either with a table of values or a graphing calculator.

7. $y = 2x^2, y = x^2, y = 4x^2$
8. $y = \frac{1}{4}x^2, y = x^2, y = 3x^2$

9. $y = 3x^2, 2y = 4x^2, y = \frac{1}{3}x^2$
10. $y = -0.75x^2, y = -4x^2, y = x^2$

11. $y = 0.5x^2, y = -3x^2, y = x^2$
12. $y = 3x^2, y = -4x^2, y = 9x^2$

13. $y = \frac{3}{2}x^2, y = \frac{2}{3}x^2, y = \frac{1}{3}x^2$
14. $y = \frac{1}{3}x^2, y = -3x^2, y = 3x^2$

15. $y = 4x^2, y = 1.5x^2, y = x^2$
16. $y = -2x^2, y = -2.5x^2, y = -4x^2$

© Prentice-Hall, Inc.

Lesson 7-2 Graphing Simple Quadratic Functions

Example

Graph the quadratic functions $y = 2x^2$ and $y = 2x^2 - 4$.

Step 1 Make a table of values.

x	$y = 2x^2$	$y = 2x^2 - 4$
-2	8	4
-1	2	-2
0	0	-4
1	2	-2
2	8	4

Step 2 Plot the points. Connect the points to form smooth curves.

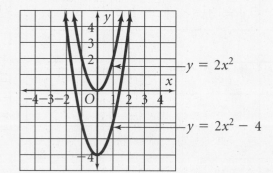

Practice

Does each function open upward or downward? Is the vertex above or below $(0, 0)$?

1. $y = x^2 + 2$ **2.** $y = 2x^2 - 3$ **3.** $3y = -9x^2 - 5$

4. $y = -x^2 + 0.05$ **5.** $\frac{1}{3}y = -6x^2 + 33$ **6.** $y = 9 + 5x^2$

Graph each quadratic function. Use a table of values or a graphing calculator.

7. $y = x^2 + 2$ **8.** $y = \frac{1}{5}x^2 + 3$ **9.** $y = x^2 + 5$

10. $y = x^2 - 6x$ **11.** $y = 1 - x^2$ **12.** $y = x^2 - 9$

13. $y = x^2 + x$ **14.** $y = -2 + \frac{1}{2}x^2$ **15.** $y = (x + 2)^2$

16. $y = 3 - x^2$ **17.** $y = 5 + x^2$ **18.** $y = 3x^2 - 4$

Solve by graphing.

19. A projectile is propelled upward. Its height h, in ft, after t seconds is given by the function $h = f(t) = 96t - 16t^2$. Graph the function. Find the maximum height reached by this projectile.

20. A school's logo is in the shape of a circle. The school wants to put a stone reproduction of the logo in a rectangular courtyard which is 15 ft by 20 ft. Find and graph the quadratic function for the area of the logo. Use $\pi = 3.14$. What is the maximum area of the logo?

Lesson 7-3 Exploring Quadratic Functions

Graphing $y = ax^2 + bx + c$

Part 1

Example

Graph the quadratic function $f(x) = x^2 - 2x - 5$.

Step 1 Find the equation of the axis of symmetry, the coordinates of the vertex, and the y-intercept.

$$x = -\frac{b}{2a} = -\frac{-2}{2(1)} = 1$$ ← **Find the equation of the axis of symmetry.**

The x-coordinate of the vertex is 1.

$$f(x) = (1)^2 - 2(1) - 5$$ ← **To find the y-coordinate, substitute 1 for x.**

$$f(x) = -6$$

The vertex is at $(1, -6)$. The value of c is -5, so the y-intercept of the graph is -5.

Step 2 Make a table of values and graph the function.

x	$f(x)$
-3	10
-2	3
-1	-2
0	-5
1	-6
3	-2
4	3

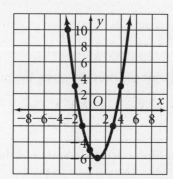

Practice

Write each equation in standard form and find the axis of symmetry.

1. $y = 4x + 2x^2 + 2$ **2.** $y - 8x = 2x^2 - 32$ **3.** $4y = 8x^2 - 32x + 4$

4. $y = 6 + 6x - 12x^2$ **5.** $\frac{1}{3}y = -4x - 2x^2$ **6.** $y = 24 - 8x^2$

Graph each quadratic function.

7. $y = x^2 + 2x + 1$ **8.** $y = 4x^2 + 2x - 2$ **9.** $f(x) = 9x^2 + 6x - 3$

10. $y = -x^2 + 4x + 10$ **11.** $y = -x^2 - x + 1$ **12.** $y = -x^2 + 3x - 4$

13. $y = x^2 + 3x - 3$ **14.** $f(x) = -x^2 - 4x$ **15.** $y = -x^2 + 3x + 4$

16. Graduates of a high school throw their mortarboards in the air at graduation. The distance traveled s, in ft, of a mortarboard after t seconds is given by the function $s = f(t) = 24t - 16t^2$. Graph the function. Find the maximum height reached by a mortarboard.

Lesson 7-3 Exploring Quadratic Functions

Quadratic Inequalities

In a graph of a linear inequality, a dashed line represents $<$ or $>$.
The curve is solid if the inequality involves $\leq$ or $\geq$.

Example

Graph the inequality $y < -0.5x^2 + 14$. Use a graphing calculator.

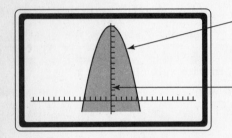

The curve is supposed to be dashed since the inequality involves $<$. A graphing calculator cannot draw a dashed line, so the curve will appear solid.

Use the calculator's **Shade** option to shade the region under the parabola.

Xmin $= -15$ Ymin $= -2$
Xmax $= 15$ Ymax $= 14$
Xscl $= 1$ Yscl $= 1$

Practice

Determine if each point satisfies the given inequality.

1. $y < 2x^2 + 2x + 3$
$(4, 26)$

2. $y < 2x^2 - 12x + 3$
$(5, 2)$

3. $y \leq -3x^2 - 3x + 6$
$(-2, 0)$

Graph each inequality. Use a graphing calculator. State whether the curve is solid or dashed.

4. $y > -x^2 + 2x - 1$

5. $y \leq 0.5x^2 - 0.25x + 2$

6. $y \geq -2x^2 + 2x$

7. $y < 3x^2 - 5x + 2$

8. $y > x^2 - x + 10$

9. $y < 5x^2 - 5$

10. $y \leq x^2 + 2x + 4$

11. $y < -x^2 - 3x - 9$

12. $y \leq 2x^2 + 2x - 2$

13. A high school is hosting an experimental car race, where all of the cars must be powered by either human power or solar power. There is a bridge over the race track where the judges sit. The bridge's shape is modeled by the quadratic function $y = -0.2x^2 + 8$. One car entered in the race has a curved top to make it more aerodynamic; its shape is modeled by the quadratic function $y = -0.15x^2 + 5$. Graph the functions and determine whether the car will fit under the bridge.

14. A train tunnel runs through a mountain. The train company is considering a new, taller locomotive and wishes to determine if it will fit through the tunnel. The tunnel's shape is modeled by the quadratic function $y = -0.15x^2 + 28$. The locomotive is 16 ft wide and 24 ft tall, and goes down the center of the tunnel. Graph the functions to determine whether the locomotive will fit through the tunnel.

© Prentice-Hall, Inc.

Lesson 7-4 Square Roots

.... **Part**
▼**1**

Finding Square Roots

Example

Simplify each expression.

a. $\sqrt{25} = 5$ ⟵ positive square root

b. $-\sqrt{64} = -8$ ⟵ negative square root

c. $\pm\sqrt{\frac{16}{25}} = \pm\frac{4}{5}$ ⟵ The square roots are $\frac{4}{5}$ and $-\frac{4}{5}$.

d. $\sqrt{0} = 0$ ⟵ There is only one square root of 0.

e. $\sqrt{-49}$ is undefined   ⟵ For real numbers, the square root of a negative number is undefined.

f. $\sqrt{2} = 1.414\,213\,56\ldots$ ⟵ irrational number

Practice

 Find a^2 for each value of a.

1. $a = -3$ **2.** $a = \frac{3}{10}$ **3.** $a = 6$

4. $a = -10$ **5.** $a = \frac{9}{5}$ **6.** $a = -\frac{5}{7}$

Simplify each expression. Round to the nearest hundredth.

7. $-\sqrt{9}$ **8.** $\sqrt{-9}$ **9.** $\pm\sqrt{16}$

10. $\sqrt{\frac{64}{81}}$ **11.** $\sqrt{7}$ **12.** $\sqrt{121}$

13. $\sqrt{0.81}$ **14.** $\sqrt{-169}$ **15.** $\sqrt{1.21}$

16. $\pm\sqrt{4}$ **17.** $\sqrt{-15}$ **18.** $-\sqrt{25}$

19. $-\sqrt{14}$ **20.** $\sqrt{36}$ **21.** $\pm\sqrt{0.25}$

22. $-\sqrt{\frac{4}{100}}$ **23.** $\pm\sqrt{0.09}$ **24.** $\sqrt{-49}$

25. $-\sqrt{49}$ **26.** $\sqrt{0.64}$ **27.** $\pm\sqrt{81}$

28. $-\sqrt{2.25}$ **29.** $\sqrt{\frac{25}{36}}$ **30.** $\pm\sqrt{169}$

31. $-\sqrt{110}$ **32.** $\sqrt{256}$ **33.** $\sqrt{-256}$

Lesson 7-4 Square Roots

Part 2

Estimating and Using Square Roots

Example 1

$\sqrt{53}$ is between what two consecutive integers?

$\sqrt{49}$ < $\sqrt{53}$ < $\sqrt{64}$ ⟵ **53 is between the two concecutive square numbers 49 and 64.**

7 < $\sqrt{53}$ < 8

$\sqrt{53}$ is between 7 and 8.

Example 2

The formula $d = \sqrt{x^2 + (2.4x)^2}$ gives the length d of the diagonal of an athletic field. Find d if $x = 50$ yd.

$d = \sqrt{x^2 + (2.4x)^2}$

$d = \sqrt{50^2 + (2.4 \cdot 50)^2}$ ⟵ **Substitute 50 for x.**

$d = \sqrt{2500 + 14,400}$ ⟵ **Simplify.**

$d = \sqrt{16,900}$

$d = 130$ ⟵ **Use a calculator.**

Practice

 What consecutive perfect square numbers would you use for estimating the following square roots?

1. $\sqrt{5}$ **2.** $\sqrt{22}$ **3.** $\sqrt{17}$

4. $\sqrt{120}$ **5.** $-\sqrt{99}$ **6.** $-\sqrt{123}$

Use a calculator to simplify each expression. Round to the nearest hundredth.

7. $\sqrt{18}$ **8.** $\sqrt{37}$ **9.** $\sqrt{98}$ **10.** $\sqrt{83}$

11. $\sqrt{125}$ **12.** $\sqrt{33}$ **13.** $\sqrt{72}$ **14.** $\sqrt{45}$

15. The area A of a pie is 53.38 in^2. The formula for the area is $A = \pi r^2$. Find r to the nearest hundredth.

16. The length of a rectangular poster is 3 times the width. The area of the rectangle is 192 in.^{2}. Find the dimensions of the poster.

Lesson 7-5 Solving Quadratic Equations

Using Square Roots to Solve Equations

Part **1**

Example

Solve $3x^2 - 24 = 0$.

$3x^2 - 24 + 24 = 0 + 24$ ⟵ **Add 24 to each side.**

$\qquad 3x^2 = 24$

$\qquad\quad x^2 = 8$ ⟵ **Divide each side by 3.**

$\qquad \sqrt{x^2} = \pm\sqrt{8}$ ⟵ **Find the square roots.**

$\qquad\quad x \approx \pm 2.83$ ⟵ **Use a calculator.**

Practice

Find each square root.

1. $\sqrt{36}$ 2. $\sqrt{64}$ 3. $\sqrt{121}$

4. $\sqrt{169}$ 5. $\sqrt{16}$ 6. $\sqrt{81}$

Solve each quadratic equation. Round solutions to the nearest hundredth.

7. $3x^2 - 27 = 0$ 8. $-7x^2 = -56$ 9. $10x^2 - 90 = 0$

10. $-4x^2 + 100 = 0$ 11. $-7x^2 = -252$ 12. $2x^2 - 14 = 0$

13. $2x^2 - \frac{1}{2} = 0$ 14. $5x^2 = 125$ 15. $x^2 - 10 = 71$

16. $4x^2 + 18 = 162$ 17. $-2x^2 + 15 = 7$ 18. $\frac{1}{2}x^2 = 128$

19. $3x^2 + 19 = 28$ 20. $\frac{1}{8}x^2 - 10 = -2$ 21. $-4x^2 + 20 = 4$

22. $2x^2 - 20 = 142$ 23. $-x^2 + 20 = -16$ 24. $9x^2 - 235 = 206$

25. $3x^2 - 12 = 0$ 26. $5x^2 - 60 = 100$ 27. $2x^2 = 72$

Write a quadratic equation and solve.

28. The length of a rectangular floor is twice the width. The area of the floor is 32 ft². What are the dimensions of the room?

29. When the square of a number is added to 6 times the number, the sum is 16. Find the numbers.

30. Find the perimeter of a square whose area is 125 cm².

31. The area of a circular pond is 2,826 m². Find the length of the radius. (Remember $A = \pi r^2$.)

32. The product of two positive integers is 36. Find the two numbers if the larger exceeds the smaller by 5.

Lesson 7-5 Solving Quadratic Equations

Finding the Number of Solutions

Example

Solve each equation by graphing the related function.

a. $x^2 + 9 = 0$
Graph $y = x^2 + 9$.

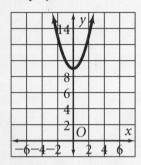

There is no solution.

b. $x^2 - 9 = 0$
Graph $y = x^2 - 9$.

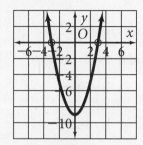

There are two solutions, $x = \pm3$.

c. $x^2 = 0$
Graph $y = x^2$.

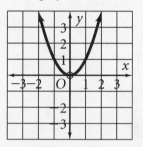

There is one solution, $x = 0$.

Practice

Find the square root of each number.

1. 36   **2.** 64   **3.** 100

4. 20   **5.** 125   **6.** 72

Solve each equation by graphing the related function. Tell the number of solutions for each equation.

7. $x^2 + 8 = 0$ **8.** $x^2 - 10 = 0$ **9.** $x^2 - 3 = -3$

10. $x^2 - 16 = 0$ **11.** $2x^2 + 12 = 12$ **12.** $3x^2 + 9 = 0$

13. $5x^2 - 125 = 0$ **14.** $-4x^2 = 16$ **15.** $6x^2 - 36 = -12$

16. $2x^2 - 25 = 103$ **17.** $-3x^2 - 18 = -45$ **18.** $7x^2 + 15 = 1$

19. $4x^2 = 1$ **20.** $2x^2 + 6 = 0$ **21.** $-4x^2 - 10 = 0$

22. $3x^2 = 0$ **23.** $9x^2 - 16 = 0$ **24.** $9x^2 + 14 = 0$

25. $2x^2 - 12 = 0$ **26.** $-x^2 = 0$ **27.** $2x^2 - 1 = 0$

28. $-2x^2 + 5 = 0$ **29.** $-3x^2 + 6 = 0$ **30.** $3x^2 - 8 = 0$

31. $5x^2 = 0$ **32.** $3x^2 - 6 = 0$ **33.** $2x^2 + 1 = 0$

34. $9x^2 - 3 = 0$ **35.** $-2x^2 - 7 = 0$ **36.** $-4x^2 + 10 = 0$

Lesson 7-6 Using the Quadratic Formula

To solve quadratic equations you may use the quadratic formula. If

$ax^2 + bx + c = 0$ and $a \neq 0$, then $x = \dfrac{-b \pm \sqrt{b^2 - 4ac}}{2a}$.

Example

Solve $2x^2 - 5x + 2 = 0$ using the quadratic formula.

$x = \dfrac{-b \pm \sqrt{b^2 - 4ac}}{2a}$ ←   **Use the quadratic formula.**

$x = \dfrac{-(-5) \pm \sqrt{(-5)^2 - 4(2)(2)}}{2(2)}$ ←   **Substitute 2 for a, −5 for b, and 2 for c.**

$x = \dfrac{5 \pm \sqrt{9}}{4}$

$x = \dfrac{5 + 3}{4}$ or $x = \dfrac{5 - 3}{4}$ ←   **Write two solutions.**

$x = 2$ or $x = \dfrac{1}{2}$

The solutions are 2 and $\frac{1}{2}$.

Practice

Put each equation into standard form and identify a, b, and c.

1. $2x - 9 = 3x^2 + 2$ **2.** $5x^2 - 3x = 10 - 6x$

3. $7 - 4x - 3x^2 = 9 - 5x$ **4.** $-x^2 + 7x = 3x - 9$

5. $x^2 - 5x = 3x + 8$ **6.** $2x^2 = 800$

Solve each quadratic equation using the quadratic formula. Round the solution to the nearest hundredth.

7. $x^2 - x - 2 = 0$ **8.** $x^2 - 6x + 9 = 0$

9. $-3x^2 + 4x - 1 = 0$ **10.** $x^2 + 6x - 3 = 0$

11. $x^2 - 2x - 3 = 0$ **12.** $2x^2 - 3x - 2 = 0$

13. $x^2 + 4x + 4 = 0$ **14.** $x^2 - 7x + 6 = 0$

15. $16x^2 - 6x - 4 = 0$ **16.** $4x^2 = -11x - 7$

17. $2x^2 - 6x + 4 = 4$ **18.** $x^2 + 4x + 4 = 0$

19. $3x^2 - 8x + 2 = 6$ **20.** $-3x^2 - 2x + 8 = 0$

21. $2x^2 - 3x - 12 = 0$ **22.** $3x^2 + 4x + 3 = 0$

23. $4x^2 - 2x + 6 = 0$ **24.** $x^2 - 3x + 2 = 0$

Lesson 7-7 Using the Discriminant

The quantity $b^2 - 4ac$ is called the **discriminant** of a quadratic equation. The discriminant is part of the quadratic formula. The value of the discriminant tells you the number of solutions.

Discriminant	# of solutions
$b^2 - 4ac > 0$	two solutions
$b^2 - 4ac = 0$	one solution
$b^2 - 4ac < 0$	no solution

Example

Find the number of solutions of $4x + 4 = -x^2$.

$$x^2 + 4x + 4 = 0 \qquad \longleftarrow \text{ Write in standard form.}$$

$$b^2 - 4ac = (4)^2 - (4)(1)(4) \qquad \longleftarrow \text{ Substitute for } a, b, \text{ and } c.$$

$$= 16 - 16$$

$$= 0$$

Since $0 = 0$, the equation has one solution.

Practice

Find the value of the discriminant $b^2 - 4ac$ for each equation.

1. $4x^2 - 4x + 1 = 0$ **2.** $2x^2 + 3x + 2 = 0$ **3.** $-x^2 + 4x - 2 = 0$

4. $x^2 + 2x - 15 = 0$ **5.** $x^2 - 5x + 4 = 0$ **6.** $x^2 - 8x + 16 = 0$

Find the number of solutions of each equation.

7. $x^2 - 10x + 25 = 0$ **8.** $x^2 + 4x - 6 = 0$ **9.** $x^2 - x - 3 = 0$

10. $x^2 + 4 = 0$ **11.** $x^2 + 4x + 3 = 0$ **12.** $x^2 - 5x + 4 = 0$

13. $x^2 + 2x - 15 = 0$ **14.** $2x^2 + 3 = 5x$ **15.** $x^2 + 3x = 3 = 0$

16. $3x^2 + 7x + 4 = 0$ **17.** $-5x^2 - 3x = -10$ **18.** $16x^2 + 1 = -8x$

19. $4x^2 = -11x - 7$ **20.** $x^2 + x + 3 = 0$ **21.** $-9x^2 + 6x - 4 = 0$

22. $3x^2 - 2x - 2 = 0$ **23.** $2x^2 + 7x - 2 = 0$ **24.** $3x^2 + 11x - 8 = 0$

25. $x^2 - 8x + 16 = 0$ **26.** $x^2 - 10x + 25 = 0$ **27.** $2x^2 - 3x + 2 = 0$

28. $3x^2 - 4x + 5 = 0$ **29.** $-2x^2 + 7x - 3 = 0$ **30.** $-6x^2 + 5x - 4 = 0$

31. $-2x^2 - 3x + 2 = 0$ **32.** $9x^2 - 30x + 25 = 0$ **33.** $4x^2 + 12x + 9 = 0$

© Prentice-Hall, Inc.

Lesson 8-1 Exploring Exponential Functions

Part 1

Exploring Exponential Patterns

Example

At the current human population growth rate of 1.8% per year, the human population doubles roughly every 39 yr. If there are 5.3 billion people today and the growth rate stays the same, how many people will there be in 156 yr?

Time	No. of People (in billions)
initial	5.3
39 yr	10.6
78 yr	21.2
117 yr	42.4
156 yr	84.8

← Use the problem-solving strategy *Make a Table.*

← To double the amount, multiply the previous period's total by 2.

← In 156 yr, there will be 84.8 billion people.

Practice

Simplify.

1. $5 \times 2 \times 2 \times 2$ 2. $10 \times 3 \times 3$ 3. $4 \times 4 \times 4 \times 4 \times 4$

4. $2 \times 5 \times 5$ 5. $6 \times 6 \times 6$ 6. $9 \times 3 \times 3 \times 3$

7. The number of students taking a foreign language in a school doubles every 4 yr. For planning purposes and to be sure the school hires enough teachers, the school needs to know how many students will be taking a foreign language in the future. If there are currently 100 students taking foreign languages and the rate of growth stays the same, how many students will be taking a foreign language in 20 yr?

8. An interstate highway is used by 4 million vehicles this year. Due to an increase in tourist attractions, the traffic is expected to double every 12 yr. How many vehicles per year will use the highway in 36 yr?

9. Because of steadily improving technology, a microchip plant is able to triple its annual production capacity every 10 yr. If the plant currently manufactures 300,000 microchips every year, how many per year will they manufacture in 20 yr?

10. A credit card company doubles the credit limit every nine months for those customers who have not been late with payments. If a customer starts with a $500 limit, what will his credit limit be in 3 yr?

Lesson 8-1 Exploring Exponential Functions

Part 2

Evaluating Exponential Functions

For all numbers a and for $b > 0$, the function $y = a \cdot b^x$ is an **exponential function.**

Example

Evaluate the exponential function
$f(x) = 3(5^x)$ for the domain $\{4, 5, 6, 7\}$.

x	$f(x) = 3(5^x)$	$f(x)$
4	$3 \cdot 5^4 = 1875$	1875
5	$3 \cdot 5^5 = 9375$	9375
6	$3 \cdot 5^6 = 46{,}875$	46,875
7	$3 \cdot 5^7 = 234{,}375$	234,375

Practice

Write each pattern in exponential form.

1. $3 \times 4 \times 4$ **2.** $8 \times 3 \times 3 \times 3$ **3.** $2 \times 2 \times 2 \times 2 \times 2$

4. $5 \times 5 \times 5$ **5.** $10 \times 6 \times 6 \times 6$ **6.** $3 \times 3 \times 3$

Evaluate each exponential function.

7. $t(n) = 2(3^n)$ for the domain $\{4, 5, 6\}$

8. $y = 3(2^x)$ for $x = 4, 6$

9. $m = 5^n$ for $n = 1, 2, 3, 4$

10. $q = 4(7^p)$ for $p = 2, 3, 4$

11. $f(x) = 8^x$ for the domain $\{3, 5\}$

12. $b = 10(2^c)$ for $c = 5, 6, 7$

13. $u = 3(4^v)$ for $v = 2, 4, 6$

14. $j = (9^k)$ for $k = (2, 3, 4)$

15. $y = (4^x)$ for $x = 2, 3, 5$

16. $f(x) = 8(7^x)$ for the domain $\{2, 3\}$

17. $t(n) = 5(4^n)$ for the domain $\{3, 5\}$

18. $h = (5^g)$ for $g = 3, 5, 7$

19. $d = 2(10^c)$ for $c = 4, 5, 6$

20. $e = 3(6^r)$ for $r = 2, 3, 4$

21. $y = (9^x)$ for $x = 2, 4$

22. $f(x) = 5(3^x)$ for the domain $\{1, 2, 4\}$

23. $y = 1^x$ for $x = 12, 23$

24. $a = 10^b$ for $b = 3, 4$

25. $c = 7(4^d)$ for $d = 1, 2$

26. $x = 11^y$ for $y = 2, 3$

27. $m = 4(3^n)$ for $n = 2, 3$

28. $x = 7(8^y)$ for $y = 2, 4$

29. $a = 4(5^b)$ for $b = 2, 3, 4$

30. $e = 6(3^f)$ for $f = 2, 4, 6$

31. $l = 2(7^m)$ for $m = 2, 4, 5$

32. $p = 3(5^q)$ for $q = 3, 5$

33. $s = 10(1^t)$ for $t = 3, 5, 7$

34. $x = 2(4^y)$ for $y = 2, 4, 6$

35. $c = 9(2^d)$ for $d = 1, 3, 5$

36. $a = 3(7^b)$ for $b = 2, 4$

37. $g = 5(6^h)$ for $h = 1, 2, 3$

38. $j = 7(7^k)$ for $k = 1, 2, 3$

39. $p = 8(2^q)$ for $q = 2, 4, 5$

40. $x = 12(3^y)$ for $y = 1, 2, 3$

Lesson 8-1 Exploring Exponential Functions

Graphing Exponential Functions

Example

Suppose an organism divides in two every hour. The function $f(t) = 2^t$ models the growth of the organism after t hours. Graph the function.

Make a table of values.

t	$f(t) = 2^t$	$f(t)$
1	$2^1 = 2$	2
2	$2^2 = 4$	4
3	$2^3 = 8$	8
4	$2^4 = 16$	16
5	$2^5 = 32$	32

Plot the points. Connect the points with a smooth curve.

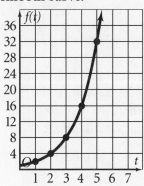

Practice

Evaluate.

1. $y = 5^3$

2. $y = (3^5)$

3. $y = 9^2$

4. $y = 1.75^2$

5. $y = 2^4$

6. $y = 2.5^3$

Make a table of values and graph the function.

7. Abby is offered a part-time job as a radio announcer. Her beginning hourly salary is $5.00. Her contract specifies that if the ratings of the show remain acceptable, she will receive a raise of 10% each year. Assuming the rankings remain acceptable, her salary growth can be modeled by the function $f(x) = 5 \cdot 1.1^x$, where x is the number of years of employment.

8. In a certain town, the number of households with access to the Internet computer network increases by 20% every six months. After the first 6 mo, 32 homes had access. This growth is modeled by the function $f(x) = 32 \cdot 1.2^x$, where x is the number of six-month periods after the first one.

9. Molly and Tong are student managers of their school store. The store takes in $400 per month of business. They plan to use aggressive advertising, which they believe will increase business 5% per month. This growth is modeled by $f(x) = 400 \cdot 1.05^x$.

10. $y = \frac{1}{2} \cdot 2^x$

11. $y = 4^x$

12. $y = \left(\frac{1}{2}\right)^x$

13. $y = \left(\frac{4}{3}\right)^x$

14. $y = 3^x$

15. $y = \frac{1}{3} \cdot 3^x$

Lesson 8-2 Exponential Growth

Modeling Exponential Growth

Part
1

You can use an exponential function in the form $y = a \cdot b^x$ to model exponential growth.

$$y = a \cdot b^x$$

starting amount →

← number of increases

the base, called the growth factor

Example

Christopher is considering opening an art gallery. Based on how similar galleries have operated, he predicts that during the first month, the gallery will have 400 visitors, and that there will be a 5% increase every month. How many visitors can he expect in the twelfth month?

Step 1 Use an exponential function to model repeated percent increases.

Relate $y = a \cdot b^x$

Define x = the number of months after the first month

y = the number of visitors during various months

b = growth factor = 100% + 5% = 105% = 1.05

a = number of visitors during the first month = 400

Write $y = 400 \cdot 1.05^x$ ← **Substitute values for the initial number a and the growth factor b.**

Step 2 There are 11 mo after the first month, so solve the equation for $x = 11$.

$y = 400 \cdot 1.05^{11}$ ← **Substitute.**

$= 684.135\ 743$

There will be about 684 visitors during the twelfth month.

Practice

Evaluate. Round to the nearest hundredth.

1. 1.03^8 **2.** $4(1.33^{13})$ **3.** $10(1.99^4)$

4. $5(1.75^4)$ **5.** $2(1.12^3)$ **6.** $3(1.25^{12})$

7. Patricia is the president of a software company. Every year her salary increases by the same percent as sales. Sales increase 5% every year. If her starting salary is $30,000/yr, what will her salary be in 7 yr?

8. A public library's membership increases 3% every year. In 1990, their membership was 275. Approximately how many members will the library have in 2007?

Lesson 8-2 Exponential Growth

Part
2

Finding Compound Interest

When a bank pays interest on both the money initially deposited *and* on the interest already earned, the bank is paying **compound interest**. Compound interest is an exponential situation.

Example

Tierra deposited $400 in a savings account when she graduated from high school. The account paid 6% annual interest, compounded semi-annually (twice a year). She withdrew the balance when she retired 47 yr later. What was the account balance?

Relate $y = a \cdot b^x$ ← **Use an exponential function.**

Define $x =$ the number of interest periods

$y =$ the balance at various times

$a = 400$ ← **Initial deposit**

$b = 1 + \frac{0.06}{2}$ ← **There are two periods in each year, so divide the interest into two parts.**

$= 1.03$

Write $y = 400 \cdot 1.03^x$

$= 400 \cdot 1.03^{94}$ ← **Since payments are made twice a year, multiply 47 · 2 to get 94 interest payments.**

$= 6438.120\ 688$

The balance after 47 yr will be $6438.12.

Practice

Evaluate. Round to the nearest hundredth.

1. 1.05^{12} 2. 1.025^{38} 3. 1.09^8

4. 1.05^4 5. 1.0325^{24} 6. 1.045^{48}

7. Kaila received a $200 prize for her science fair project when she was in eighth grade. She chose to put the money in a savings account to help pay for college. The account paid 4% annual interest, compounded quarterly. What is the balance of the account 4 yr later?

8. Milton was hired shortly after graduating from high school. His new employer, a law firm, paid him a $750 "signing bonus." He put this money in an account that paid 5% interest, compounded semi-annually. What is the balance of his account at the end of 5 yr?

9. A movie theatre's matinee ticket price increases 4% every year. The price of a ticket is currently $3.50. How much will it cost in 10 yr?

Lesson 8-3 Exponential Decay

When b is between 0 and 1 in the exponential function $y = a \cdot b^x$, the function can model **exponential decay**, and b is called the **decay factor**.

Example

Suppose an endangered species of salamanders are threatened by increased development, which started in 1985. Graph the function $y = 2750 \cdot 0.915^x$, where y is the population of salamanders and x is the number of years since 1985. Predict the salamander population in 1993.

Make a table of values.
$y = 2750 \cdot 0.915^x$

x	y
0	2750
5	1764
10	1131
15	726
20	465

Graph the points.
Draw a smooth curve through the points.

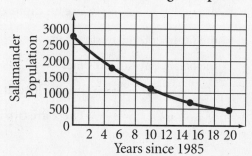

The y-value of 1,350 corresponds to the x-value of 8.

$$y = 2750 \cdot 0.915^8$$

⟵ **Use the equation to predict the population 8 yr later and to check the answer from the graph.**

$$= 1351.143\,407$$

There were about 1351 salamanders in 1993.

Practice

Mini ▶ Help

Evaluate. Round to the nearest hundredth.

1. $350 \cdot 0.805^{12}$ **2.** $2025 \cdot 0.995^{18}$ **3.** $80 \cdot 0.89^5$

4. $9250 \cdot 0.55^{14}$ **5.** $208 \cdot 0.9325^4$ **6.** $347 \cdot 0.745^{10}$

7. Suppose that attendance at a minor league baseball team's games has been declining for a number of years. Graph the function $y = 5750 \cdot 0.975^x$, where y is the attendance at a game and x is the number of years since the decline began. Use the equation to predict attendance at a game 15 yr after the decline started.

8. Suppose that the number of minutes of music videos per hour shown on a music television channel has dropped every year since 1990, being replaced by commercials and non-music programming. Graph the function $y = 50 \cdot 0.9125^x$, where y is the number of minutes of music videos per hour and x is the number of years since 1990. Use the equation to predict the minutes of music videos shown per hour in 1998.

© Prentice-Hall, Inc.

Lesson 8-4 Zero and Negative Exponents

Part 1

Using Zero and Negative Integers as Exponents

For any nonzero number a, $a^0 = 1$. For any nonzero number a and any integer n, $a-n = \frac{1}{a^n}$.

Example

Simplify each expression.

a. 3^{-4}

$3^{-4} = \frac{1}{3^4}$ ⟵ definition of a negative exponent

$= \frac{1}{81}$

b. 9^0

$9^0 = 1$ ⟵ definition of a zero exponent

c. 5^{-3}

$5^{-3} = \frac{1}{5^3}$ ⟵ definition of a negative exponent

$= \frac{1}{125}$

Practice

Evaluate.

1. $\frac{1}{2^7}$

2. $\frac{1}{4^4}$

3. $\frac{1}{3^6}$

4. $\frac{1}{10^3}$

5. $\frac{1}{(-5)^2}$

6. $\frac{1}{8^0}$

Simplify each expression, writing each as a simple fraction or as an integer, as appropriate.

7. 3^{-4} **8.** 4^{-2} **9.** $(-15)^0$

10. 6^{-3} **11.** 8^{-2} **12.** 11^{-3}

13. 2^{-10} **14.** 8^0 **15.** 5^{-4}

16. $(-3)^{-3}$ **17.** 5^0 **18.** 12^{-2}

19. 9^{-4} **20.** $(-13)^{-2}$ **21.** 8^{-3}

22. $(-4)^{-5}$ **23.** 7^{-6} **24.** $(-2)^{-9}$

25. $(-2)^{-10}$ **26.** 14^{-2} **27.** 10^{-5}

28. $(-16)^{-2}$ **29.** $(-5)^0$ **30.** 4^{-4}

31. 3^{-11} **32.** $(-5)^{-3}$ **33.** 9^0

Lesson 8-4 Zero and Negative Exponents

Part 2

Relating the Properties to Exponential Functions

Knowing the definitions of zero and negative integer exponents helps you analyze a more complete graph of an exponential function.

Example

Step 1 Using a graphing calculator, graph the functions $y = 3^x$ and $y = \left(\frac{1}{3}\right)^x$ on the same set of axes. Show the functions over the domain $\{-3 \le x \le 3\}$.

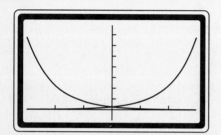

Step 2 Make a table of the values, either by using the table key on a graphing calculator or by manual calculation. Notice that for any x, 3^x and $\left(\frac{1}{3}\right)^x$ are reciprocals.

x	3^x	$\left(\frac{1}{3}\right)^x$
-3	$\frac{1}{27}$	27
-2	$\frac{1}{9}$	9
-1	$\frac{1}{3}$	3
0	1	1
1	3	$\frac{1}{3}$
2	9	$\frac{1}{9}$
3	27	$\frac{1}{27}$

Practice

Evaluate.

1. 5^2

2. $\left(\frac{1}{3}\right)^3$

3. 6^{-3}

4. $\left(\frac{1}{4}\right)^{-4}$

5. 7^{-3}

6. 11^{-3}

Using a graphing calculator, graph each pair of functions on the same set of axes over the given domain.

7. $\left(\frac{1}{4}\right)^x, 4^x, \{-2 \le x \le 2\}$

8. $\left(\frac{3}{4}\right)^x, \left(\frac{4}{3}\right)^x, \{-4 \le x \le 4\}$

9. $\left(\frac{1}{5}\right)^x, 5^x, \{-2 \le x \le 2\}$

10. $\left(\frac{3}{2}\right)^x, \left(\frac{2}{3}\right)^x, \{-4 \le x \le 4\}$

11. $\left(\frac{1}{2}\right)^x, 2^x, \{-4 \le x \le 4\}$

12. $\left(\frac{1}{3}\right)^x, 3^x, \{-2 \le x \le 2\}$

Lesson 8-5 Scientific Notation

Part 1

Writing Numbers in Scientific Notation

A number is in **scientific notation** if it is written in the form $a \times 10^n$, where n is an integer and $1 \le a \le 10$. Scientific notation can be helpful when writing very large or very small numbers.

Example

Write each number in standard notation.

a. area of the Earth's surface: $5.100\ 501 \times 10^8$ km^2

b. density of living matter in the oceans: 1.5×10^{-7} g/cm^3

a. $5.100\ 501 \times 10^8 = 5.10050100.$ ← **A positive exponent indicates a large number. Move the decimal point 8 places to the right.**

   $= 510{,}050{,}100$

b. $1.5 \times 10^{-7} = 0.0000001.5$ ← **A negative exponent indicates a small number. Move the decimal point 7 places to the left.**

   $= 0.000\ 000\ 15$

Practice

Determine if each number is in scientific notation; if it is, indicate a and n.

1. 10×5^2 **2.** 3.348×10^{18} **3.** $6 \times 10^{-3.54}$

4. 9.9325×10^2 **5.** 7.25×10^{-3} **6.** 10^{-23}

7. Make a table of the data below. Rank the planets from lowest mass to highest mass, then write each mass in standard notation.

Planet	Approximate Mass (kg)
Mercury	3.1881×10^{23}
Venus	4.883×10^{24}
Earth	5.979×10^{24}
Mars	6.418×10^{23}
Jupiter	1.901×10^{27}
Saturn	5.684×10^{26}
Uranus	8.682×10^{25}
Neptune	1.027×10^{26}
Pluto	1.08×10^{24}

Name_____ Class_____ Date_____

Lesson 8-5 Scientific Notation
• •
Calculating With Scientific Notation

Examples

Simplify $6 \times (7 \times 10^8)$. Give your answer in scientific notation.

$$6 \times (7 \times 10^8) = 42 \times 10^8 \qquad \longleftarrow \textbf{Multiply whole numbers.}$$
$$= 4.2 \times 10 \times 10^8 \qquad \longleftarrow \textbf{Write 42 as 4.2} \times \textbf{10.}$$
$$= 4.2 \times 10^9 \qquad \longleftarrow \textbf{Combine powers of 10.}$$

Simplify $\frac{7,000,000,000}{260,000,000}$. Give your answer in scientific notation.

$$\frac{7,000,000,000}{260,000,000} = \frac{7 \times 10^9}{2.6 \times 10^8} \qquad \longleftarrow \textbf{Write in scientific notation.}$$

$$= 7 \; \boxed{EE} \; 9 \; \boxed{\div} \; 2.6 \; \boxed{EE} \; 8 \; \boxed{ENTER} \qquad \longleftarrow \textbf{Use a calculator.}$$

$$= 26.923\,076\,92$$

$$= 2.692\,307\,692 \times 10^1 \qquad \longleftarrow \textbf{Write in scientific notation.}$$

Practice

 Write each expression as a single power of ten.

1. 10×10^2 **2.** $10^{18} \times 10$ **3.** $10^2 \times 10^{13}$

4. 100×10^7 **5.** 10×10^{32} **6.** $10^7 \times 10^{23}$

Simplify. Calculate each product or quotient. Give your answer in scientific notation.

7. $4 \times (2 \times 10^7)$ **8.** $9 \times (7 \times 10^{12})$ **9.** $2 \times (8 \times 10^8)$

10. $(7 \times 10^8) \div 3.5$ **11.** $(9 \times 10^{12}) \div 3$ **12.** $(2 \times 10^{11}) \div (4.75 \times 10^6)$

13. $12 \times (12 \times 10^7)$ **14.** $8 \times (7 \times 10^{13})$ **15.** $7 \times (5 \times 10^8)$

16. $(8.4 \times 10^{38}) \div 4$ **17.** $(1.75 \times 10^{12}) \div 0.25$ **18.** $(9.99 \times 10^{22}) \div (3 \times 10^{17})$

19. $25 \times (4 \times 10^7)$ **20.** $6 \times (7.25 \times 10^{26})$ **21.** $(5.675 \times 10^9) \div 1.135$

22. $3.14 \times (2.85 \times 10^6)$ **23.** $(8.25 \times 10^{19}) \div 1.25$ **24.** $(7.5 \times 10^{12}) \div (2.5 \times 10^7)$

25. $2.85 \times (4.5 \times 10^5)$ **26.** $40 \times (1.5 \times 10^{16})$ **27.** $15 \times (5.5 \times 10^{62})$

28. Sometimes scientists use the mass of the Earth as the unit for describing the mass of other astronomical bodies. One Earth Mass is approximately 5.979×10^{24} kg. Jupiter's mass is approximately 318 Earth Masses. Calculate Jupiter's mass in kg by calculating $318 \times (5.979 \times 10^{24})$.

29. The Sun's mass is approximately 332,999 Earth Masses. Calculate its mass in kg by calculating $332,999 \times (5.979 \times 10^{24})$.

30. Saturn's mass is approximately 5.684×10^{26} kg. Find Saturn's mass in Earth Masses by calculating $(5.684 \times 10^{26}) \div (5.979 \times 10^{24})$.

Lesson 8-6 A Multiplication Property of Exponents   Part 1

Multiplying Powers

For any number a and any integers m and n, $am \cdot an = am + n$.
In other words, when you multiply powers with the same base, you add
the exponents.

Example

Simplify each expression.

$p^8 \cdot q^4 \cdot p^2$

$\qquad = p^8 \cdot p^2 \cdot q^4$ ⟵ **Regroup factors.**

$\qquad = p^{8 \, + \, 2} \cdot q^4$ ⟵ **Add exponents of powers with the same base.**

$\qquad = p^{10}q^4$

$7c^3 \cdot 5d^3 \cdot 4c$

$\qquad = (7 \cdot 5 \cdot 4)(c^3 \cdot c)(d^3)$ ⟵ **Regroup factors.**

$\qquad = (140)(c^3 \cdot c^1)(d^3)$ ⟵ **Multiply coefficients.**

$\qquad = (140)(c^{3 \, + \, 1})(d^3)$ ⟵ **Add exponents of powers with the same base.**

$\qquad = 140c^4d^3$

Practice

Rewrite each expression using exponents.

1. $(x - 7)(x - 7)(x - 7)(x - 7)(x - 7)$ **2.** $y \cdot y \cdot y \cdot y \cdot y \cdot y \cdot y \cdot y$

3. $9 \cdot 9 \cdot 9 \cdot 9 \cdot v \cdot v \cdot v$ **4.** $2 \cdot 2 \cdot j \cdot j \cdot k \cdot k \cdot k \cdot k \cdot k$

5. $s \cdot s \cdot s \cdot 4 \cdot 4 \cdot 4 \cdot t \cdot t$ **6.** $z \cdot z \cdot z \cdot z \cdot z \cdot (x + 3)(x + 3)$

Simplify each expression.

7. $c^3 \cdot 2d^3 \cdot 4c^4$ **8.** $2a^8 \cdot 7b \cdot 8a^{12}$ **9.** $5e \cdot 5e^{12} \cdot 2f^6$

10. $3h^4 \cdot 7g^2 \cdot 2h^6$ **11.** $7j^6 \cdot 7k^3 \cdot 5k^3$ **12.** $l^3 \cdot 7m^{11} \cdot 9l^5$

13. $5m^5 \cdot 5n^3 \cdot 3m^6$ **14.** $5q^4 \cdot 3p^2 \cdot 3p^{11}$ **15.** $3r^6 \cdot 5s^{10} \cdot 8r^4$

16. $7u^6 \cdot 3u^4 \cdot t^{12}$ **17.** $3v^2 \cdot 4w^5 \cdot 8v^4$ **18.** $x^9 \cdot 3x^9 \cdot 7y^3$

19. $9e^3 \cdot z^5 \cdot 2e^{14}$ **20.** $a^3 \cdot 7a^4 \cdot 4b^{10}$ **21.** $2c^{12} \cdot d^8 \cdot 5c^4$

22. $4f^8 \cdot 6e^6 \cdot 3f^8$ **23.** $6h^5 \cdot 7g^7 \cdot 2h^5$ **24.** $3i^2 \cdot 6j^7 \cdot 3i^5$

25. $8k^{10} \cdot 10l^7 \cdot 4k^{10}$ **26.** $10n^7 \cdot m^6 \cdot 9m^9$ **27.** $4o^5 \cdot 10p^6 \cdot 2o^6$

28. $10q^{13} \cdot 2q^8 \cdot 5r^9$ **29.** $2t^9 \cdot 8s^9 \cdot t^6$ **30.** $5u^8 \cdot 2u^5 \cdot 4v^7$

31. $6w^2 \cdot 4x^9 \cdot 6x^{14}$ **32.** $4y^{11} \cdot 3x^8 \cdot 7x^8$ **33.** $6a^{11} \cdot 4b^4 \cdot 6a^5$

Lesson 8-6 A Multiplication Property of Exponents Part 2
Working with Scientific Notation

The property of multiplying powers with the same base can make working with scientific notation easier. Also, the property works with negative exponents.

Example

Simplify $(9 \times 10^{11})(3 \times 10^{-15})$. Give the answer in scientific notation.

$(9 \times 10^{11})(3 \times 10^{-15})$

$= (9 \times 3)(10^{11} \times 10^{-15})$ ⟵ **Regroup factors.**

$= 27 \times 10^{11 + (-15)}$ ⟵ **Multiply.**

$= 27 \times 10^{-4}$ ⟵ **Add exponents.**

$= 2.7 \times 10^1 \times 10^{-4}$ ⟵ **Write 27 as 2.7×10^1.**

$= 2.7 \times 10^{-3}$ ⟵ **Add exponents.**

Practice

Rewrite each expression in scientific notation.

1. 140,000   **2.** 0.000 000 356   **3.** 423,000,000

4. 29   **5.** 0.0936   **6.** 81.5

Simplify each expression. Give the answer in scientific notation.

7. $(3.75 \times 10^4)(5 \times 10^1)$

8. $(9.5 \times 10^{-8})(4 \times 10^{18})$

9. $(7.5 \times 10^5)(3 \times 10^{-14})$

10. $(1.5 \times 10^{11})(3 \times 10^{-12})$

11. $(4 \times 10^{-3})(7.75 \times 10^{12})$

12. $(2.95 \times 10^3)(2 \times 10^9)$

13. $(2.75 \times 10^{12})(5 \times 10^{10})$

14. $(5.75 \times 10^{-8})(8 \times 10^{-5})$

15. $(1.335 \times 10^{-32})(6 \times 10^{10})$

16. $(6 \times 10^8)(2.75 \times 10^{-7})$

17. $(8.2 \times 10^4)(4 \times 10^6)$

18. $(3.5 \times 10^7)(3.75 \times 10^{10})$

19. $(3.5 \times 10^9)(3.5 \times 10^{-18})$

20. $(2 \times 10^{22})(2.5 \times 10^{30})$

21. $(9.75 \times 10^{14})(5.25 \times 10^2)$

22. $(1 \times 10^{18})(3.9 \times 10^{-6})$

23. Each 3.5-in. floppy disk can store about 1.3 megabytes (1.3×10^4 bytes) of information. A computer lab uses these disks to back up student papers. They own 1,500 disks (1.5×10^3 disks). Calculate the disks' total capacity by simplifying $(1.3 \times 10^4)(1.5 \times 10^3)$.

24. The density of living matter in the ocean is approximately 1.5×10^{-9} g/cm^3. A scientist collects a sample of ocean water in a cube with a total volume of 3.375×10^6 cm^3. Calculate the approximate number of grams of living matter that will be in the sample by simplifying $(1.5 \times 10^{-9})(3.375 \times 10^6)$.

Lesson 8-7
More Multiplication Properties of Exponents

Part 1

Raising a Power to a Power

For any number $a \geq 0$ and any integers m and n, $(a^m)^n = a^{mn}$.

Example

Simplify $c^3(c^4)^2$.

$= c^3 \cdot c^{4 \cdot 2}$ ← **Multiply exponents in $(c^4)^2$.**

$= c^3 \cdot c^8$ ← **Simplify.**

$= c^{3+8}$ ← **Add exponents when multiplying powers with the same base.**

$= c^{11}$

Practice

 Rewrite each expression with one exponent.

1. $4^3 \cdot 4^3 \cdot 4^3 \cdot 4^3$
2. $3^6 \cdot 3^6 \cdot 3^6$
3. $5^4 \cdot 5^4$
4. $24^2 \cdot 24^2 \cdot 24^2 \cdot 24^2$
5. $9^3 \cdot 9^3 \cdot 9^3 \cdot 9^3 \cdot 9^3$
6. $15^5 \cdot 15^5 \cdot 15^5 \cdot 15^5$

Simplify each expression.

7. $a^2(a^3)^4$
8. $b^9(b^2)^{10}$
9. $c^2(c^3)^5$
10. $d^3(d^2)^{20}$
11. $c^7(c^9)^3$
12. $f^6(f^6)^6$
13. $d^7(d^3)^3$
14. $c^5(c^4)^8$
15. $g^5(g^9)^7$
16. $e^5(e^7)^2$
17. $d^3(d^5)^7$
18. $h^4(h^{12})^8$
19. $g^8(g^5)^4$
20. $e(e^7)^5$
21. $i^3(i^{11})^9$
22. $h^{10}(h^6)^5$
23. $f^2(f^6)^6$
24. $j^2(j^8)^{10}$
25. $i^9(i^3)^{10}$
26. $g^4(g^{11})^3$
27. $k(k^5)^{11}$
28. $j^7(j^4)^3$
29. $h^6(h^4)^{12}$
30. $l^2(l^2)^3$
31. $k^5(k^9)^9$
32. $i^8(i^4)^{13}$
33. $m^3(m^3)^4$
34. $l^3(l^2)^3$
35. $j^{10}(j^2)^2$
36. $n^4(n^{10})^6$
37. $m^7(m^7)^6$
38. $k^9(k^3)^3$
39. $o^5(o^7)^{10}$
40. $n^7(n^4)^5$
41. $l(l^3)^7$
42. $p^6(p^4)^9$
43. $o^1(o^2)^3$
44. $m^8(m^5)^{10}$
45. $q^7(q^{13})^7$
46. $p^8(p^5)^4$
47. $n^2(n^7)^2$
48. $r^8(r^{10})^5$
49. $q^5(q^6)^8$
50. $o^7(o^3)^{15}$
51. $s^9(s^7)^3$
52. $r^{13}(r^8)^{10}$
53. $p^3(p^2)^{17}$
54. $t^{10}(t^4)^2$
55. $s^4(s^5)^7$
56. $q^6(q^{10})^{12}$
57. $u^{11}(u^{14})^2$

Lesson 8-7
More Multiplication Properties of Exponents
Part 2

Raising a Product to a Power

For any number a and b and any integer n, $(ab)^n = a^n b^n$.

Example

Simplify $4(3c^3)^3$.

$$= 4 \cdot 3^3 (c^3)^3 \quad \longleftarrow \quad \textbf{Raise each factor to the 3rd power.}$$

$$= 4 \cdot 3^3 c^9 \quad \longleftarrow \quad \textbf{Multiply exponents.}$$

$$= 4 \cdot 27 c^9 \quad \longleftarrow \quad \textbf{Simplify.}$$

$$= 108 c^9$$

Practice

Simplify each expression.

1. $(a^3)^4$　　　　　　**2.** $(b^2)^{10}$　　　　　　**3.** $(c^3)^5$

4. $(d^2)^{20}$　　　　　　**5.** $(c^9)^3$　　　　　　**6.** $(g^9)^7$

Simplify each expression.

7. $2(9a^7)^2$　　　　　　**8.** $(7o^{10})^2$　　　　　　**9.** $2(2c^2)^6$

10. $4(7b^{11})^3$　　　　　**11.** $3(2p^5)^6$　　　　　**12.** $10(3d^4)^5$

13. $7(3c^{12})^4$　　　　　**14.** $5(4q^2)^4$　　　　　**15.** $(4e^6)^4$

16. $11(2d^3)^5$　　　　　**17.** $7(6r^3)^3$　　　　　**18.** $3(5f^8)^3$

19. $10(4e^5)^4$　　　　　**20.** $2(s^5)^5$　　　　　**21.** $4(6g^{10})^2$

22. $(5x^7)^3$　　　　　　**23.** $4(5t^4)^3$　　　　　**24.** $8(5h^{12})^3$

25. $8(12g^9)^2$　　　　　**26.** $6(13u^3)^2$　　　　　**27.** $(2i^{14})^4$

28. $7(8h^{11})^3$　　　　　**29.** $(4v^2)^4$　　　　　**30.** $5(2j^{11})^5$

31. $(6i^{13})^3$　　　　　　**32.** $10(10w^7)^3$　　　　**33.** $4(3k^5)^4$

34. $5(3j^{15})^4$　　　　　**35.** $9(9x^3)^2$　　　　　**36.** $5(8l^9)^3$

37. $3(11k^{12})^2$　　　　**38.** $(y^5)^5$　　　　　　**39.** $(12m^7)^2$

40. $4(3l^8)^3$　　　　　　**41.** $7(5z^6)^2$　　　　　**42.** $6(10n^2)^3$

43. $(10m^6)^6$　　　　　**44.** $6(5a^9)^3$　　　　　**45.** $2(9o^6)^2$

46. $2(8n^4)^2$　　　　　　**47.** $5(15b^{10})^2$　　　　**48.** $(2p^5)^3$

49. A particle accelerator is built for physics experiments. The main part of the structure is an underground tunnel that runs in a circle with a radius of 4×10^3 m. Find the approximate area of the circle by simplifying $3.14 \times (4 \times 10^3)^2$.

Lesson 8-8 Division Properties of Exponents

Part
1

Dividing Powers with the Same Base

For any number a and any integers m and n, $\dfrac{a^m}{a^n} = a^{m-n}$.

Example

Simplify $\dfrac{a^4 b^{-2}}{a^2 b^3}$. Use only positive exponents.

$$\dfrac{a^4 b^{-2}}{a^2 b^3} = a^{4-2} \cdot b^{-2-3} \qquad \longleftarrow \quad \textbf{Subtract exponents when dividing powers with the same base.}$$

$$= a^2 b^{-5}$$

$$= \dfrac{a^2}{b^5} \qquad \longleftarrow \quad \textbf{Rewrite using positive exponents.}$$

Practice

Rewrite each expression using only positive exponents.

1. a^{-3} **2.** $\dfrac{1}{b^{-2}}$ **3.** $c^3 c^{-5}$

4. d^{-20} **5.** c^{-4} **6.** $\dfrac{1}{g^{-7}}$

Simplify each expression. Use only positive exponents.

7. $\dfrac{ab^{12}}{a^2 b^4}$ **8.** $\dfrac{c^6 d^2}{c^5 d}$ **9.** $\dfrac{e^5 f^{-3}}{e^3 f^{-13}}$

10. $\dfrac{g^4}{g^{13}}$ **11.** $\dfrac{h^{24}}{h^{22}}$ **12.** $\dfrac{i^{-9} j^{-2}}{i^{-2} j^4}$

13. $\dfrac{k^{32} l^{-13}}{k^{22} l^{-9}}$ **14.** $\dfrac{m^{15} n^{21}}{m^{-2} n^{-20}}$ **15.** $\dfrac{o^{46} p^{-70}}{o^{-5} p^{30}}$

16. A certain city has a population of 4.8×10^5. In one year, 1.2×10^6 gal of ice cream are sold. Find the average ice cream consumption in gallons per person by finding the quotient $\dfrac{1.2 \times 10^6}{4.8 \times 10^5}$.

17. A large company employs 38,000 people and wishes to determine the average amount of time each employee spends on the phone per month. The average total phone usage per month is 1.9×10^5 h. Calculate the average number of hours spent on the phone by each employee by calculating the quotient $\dfrac{1.9 \times 10^5}{3.8 \times 10^4}$.

Lesson 8-8 Division Properties of Exponents

Raising a Quotient to a Power

For any number a, any nonzero number b, and any integer n, $\left(\dfrac{a}{b}\right)^n = \dfrac{a^n}{b^n}$.

Example

Simplify each expression.

a. $\left(-\dfrac{3}{4}\right)^{-3} = \left(\dfrac{-3}{4}\right)^{-3}$ ⟵ **Write the fraction with a negative numerator.**

$\qquad = \dfrac{(-3)^{-3}}{4^{-3}}$ ⟵ **Raise the numerator and denominator to the −3rd power.**

$\qquad = \dfrac{4^3}{(-3)^3}$ ⟵ **Apply the definition of negative exponents.**

$\qquad = -\dfrac{64}{27}$ ⟵ **Simplify.**

b. $\left(\dfrac{2x^2}{3}\right)^3 = \dfrac{(2x^2)^3}{3^3}$ ⟵ **Raise the numerator and denominator to the 3rd power.**

$\qquad = \dfrac{2^3(x^2)^3}{3^3}$ ⟵ **Apply a multiplication property of exponents.**

$\qquad = \dfrac{8x^6}{27}$ ⟵ **Multiply exponents and simplify.**

Practice

Simplify each expression.

1. $\dfrac{(-2)^3}{3^3}$ **2.** $\dfrac{4^3(f^2)^3}{(-2)^3}$ **3.** $\dfrac{4^2k^2}{(-4)^2}$

Simplify each expression.

4. $\left(\dfrac{2}{3}\right)^3$ **5.** $\left(\dfrac{3y}{2y^2}\right)^3$ **6.** $\left(-\dfrac{x^4}{2x}\right)^4$

7. $\left(\dfrac{7n^5}{8n^3}\right)^2$ **8.** $\left(-\dfrac{3}{7}\right)^2$ **9.** $\left(\dfrac{2p^3}{3p}\right)^3$

10. $\left(\dfrac{1}{3g^3}\right)^3$ **11.** $\left(\dfrac{10r^5}{2r^4}\right)^3$ **12.** $\left(\dfrac{v^3}{12}\right)^2$

13. $\left(-\dfrac{3e^7}{6e}\right)^2$ **14.** $\left(-\dfrac{4}{5}\right)^3$ **15.** $\left(\dfrac{11t^{12}}{12t^{13}}\right)^2$

Lesson 9-1 The Pythagorean Theorem

Solving Equations Using the Pythagorean Theorem

Part 1

The Pythagorean Theorem:
In a right triangle, the sum of the squares of the lengths of the legs is equal to the square of the length of the hypotenuse. $a^2 + b^2 = c^2$

Example

A right triangle has legs of lengths 6 cm and 8 cm. What is the length of the hypotenuse?

$$a^2 + b^2 = c^2 \quad \longleftarrow \text{ Use the Pythagorean theorem.}$$

$$6^2 + 8^2 = c^2 \quad \longleftarrow \text{ Substitute 6 for } a \text{ and 8 for } b.$$

$$36 + 64 = c^2 \quad \longleftarrow \text{ Simplify.}$$

$$\sqrt{100} = \sqrt{c^2} \quad \longleftarrow \text{ Take the square root of each side.}$$

$$c = 10$$

The hypotenuse has length 10 cm.

Practice

Use a calculator to solve.

1. $x^2 = 9$ 2. $x^2 = 15$ 3. $x^2 = 10$

4. $x^2 = 12$ 5. $x^2 = 8$ 6. $x^2 = 16$

Use the triangle at the right. Find the length of the missing side to the nearest tenth. Use a calculator.

7. $a = 10, b = \blacksquare, c = 15$ 8. $a = 12, b = 16, c = \blacksquare$

9. $a = 2, b = 3, c = \blacksquare$ 10. $a = 5, b = 6, c = \blacksquare$

11. $a = 4, b = 5, c = \blacksquare$ 12. $a = 3, b = 5, c = \blacksquare$

13. $a = \blacksquare, b = 4, c = 10$ 14. $a = 3, b = 4, c = \blacksquare$

Solve.

15. A carpenter braces an 8 ft by 15 ft wall by nailing a board diagonally across the wall. How long is the bracing board?

16. The diagonal of a square is $8\sqrt{2}$ cm. Find the length of a side of the square.

17. The lengths of the sides of a right triangle are given by three consecutive integers. Find the lengths of the sides.

18. A wire is stretched from the top of a 4-ft pole to the top of a 9-ft fence. If the pole and fence are 12 ft apart, how long is the wire?

19. A baseball diamond is also a square 90 ft on a side. How far is it from second base to home plate?

© Prentice-Hall, Inc.

Lesson 9-1 The Pythagorean Theorem

Using the Converse

To find out whether a triangle is a right triangle, use the converse of the Pythagorean theorem.

The Converse of the Pythagorean Theorem:
If a triangle has sides of lengths a, b, c, and $a^2 + b^2 = c^2$, then the triangle is a right triangle with hypotenuse of length c.

Example

A triangle has sides of length 6, 8, 10. Is it a right triangle?

Apply the converse of the Pythagorean theorem.

$$a^2 + b^2 \stackrel{?}{=} c^2$$

$$6^2 + 8^2 \stackrel{?}{=} 10^2 \qquad \longleftarrow \text{ Substitute 10 for } c, \text{ since 10 is the length of the longest side.}$$
$$\text{Substitute 6 and 8 for } a \text{ and } b.$$

$$36 + 64 \stackrel{?}{=} 100$$

$$100 = 100 \ ✔$$

The triangle is a right triangle.

Practice

 Find each value.

1. $\sqrt{3^2 + 4^2}$ **2.** $\sqrt{3^2 + 5^2}$ **3.** $\sqrt{6^2 + 8^2}$

4. $\sqrt{1^2 + 2^2}$ **5.** $\sqrt{9^2 + 12^2}$ **6.** $\sqrt{5^2 + 6^2}$

Can each set of 3 numbers represent the lengths of the sides of a right triangle? Explain your answer.

7. 5, 7, 10	**8.** 7, 9, 11	**9.** 3, 8, 12
10. 4, 7, 12	**11.** 12, 16, 20	**12.** 6, 8, 10
13. 3, 4, 5	**14.** 13, 14, 15	**15.** 15, 20, 25
16. 6, 8, 14	**17.** 9, 12, 15	**18.** 4, 6, 8
19. 1, 2, 3	**20.** 7, 8, 9	**21.** 18, 24, 30
22. 5, 12, 13	**23.** 2, 6, 10	**24.** 3, 7, 11
25. 1.5, 3, 5	**26.** 4.5, 6, 7.5	**27.** 5, 7, 9.5
28. 2.5, 6, 6.5	**29.** 3.5, 4.5, 5.5	**30.** 10.5, 12, 14
31. 5, 6.5, 7	**32.** 11, 12, 13.5	**33.** 14, 15, 17.5
34. 5, 10, 15	**35.** 7.5, 10, 25	**36.** 10, 20, 30
37. 10.5, 14, 17.5	**38.** 5, 15, 25	**39.** 24, 25, 26
40. 21, 28, 35	**41.** 4.5, 5, 7	**42.** 8, 16, 24

Lesson 9-2 The Distance Formula

Part
1

Finding the Distance

The distance d between any two points (x_1, y_1) and (x_2, y_2) is
$d = \sqrt{(x_2 - x_1)^2 + (y_2 - y_1)^2}$.

Example

Find AB for points $A(0, 0)$ and $B(3, 7)$.

$$d = \sqrt{(x_2 - x_1)^2 + (y_2 - y_1)^2}$$

$$AB = \sqrt{(3 - 0)^2 + (7 - 0)^2}$$

$$AB = \sqrt{3^2 + 7^2}$$

$$AB = \sqrt{58}$$

$$AB \approx 7.62$$

Practice

 Solve.

1. $\sqrt{x} = 4$ 2. $\sqrt{x} = 5$ 3. $\sqrt{x} = 7$

4. $\sqrt{x} = 9$ 5. $\sqrt{x} = 1$ 6. $\sqrt{x} = 2$

Find the distance between each pair of points. Round your answer to the nearest hundredth.

7. $(3, 4), (-2, -1)$ 8. $(-1, 2), (3, 4)$ 9. $(-1, 0), (-2, 0)$

10. $(0, 0), (4, 5)$ 11. $(0, 3), (4, 5)$ 12. $(0, 1), (5, 0)$

13. $(-1, 1), (-2, 2)$ 14. $(5, 7), (0, 2)$ 15. $(1, 0), (3, 0)$

16. $(6, 3), (3, 4)$ 17. $(12, 10), (11, 9)$ 18. $(5, 6), (7, 8)$

19. $(-5, -7), (-6, -8)$ 20. $(1,2), (3,4)$ 21. $(2, 4), (3, 5)$

22. $(6, 2), (3, 4)$ 23. $(0, 0), (-5, -2)$ 24. $(6, 5), (5, 6)$

25. $(0, -2), (7, 3)$ 26. $(8, 20), (7, 21)$ 27. $(1, 3), (0, 0)$

28. $(2, -3), (4, -1)$ 29. $(0, 0), (-1, 1)$ 30. $(1, -1), (-1, 1)$

31. $(8, 2), (6, 2)$ 32. $(-5, -1), (0, 0)$ 33. $(0, 0), (4, 7)$

34. In a football game, a quarterback throws a pass from the 5-yd line, 10 yd from the sideline. The pass is caught by a wide receiver on the 45-yd line, 40 yd from the same sideline. How long was the pass?

35. Suppose you are standing at a point that is 10 m from the base of Minnehaha Falls. Minnehaha Falls is 12.5 m high. How long is the line from your point to the top of the falls?

Lesson 9-2 The Distance Formula

Part 2

Using the Midpoint Formula

The Midpoint Formula

The midpoint M of a line segment with endpoints $A(x_1, y_1)$ and $B(x_2, y_2)$ is $\left(\dfrac{x_1 + x_2}{2}, \dfrac{y_1 + y_2}{2}\right)$.

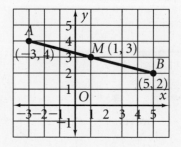

Example

Find the midpoint of the segment from $A(-3, 4)$ to $B(5, 2)$.

$$\left(\frac{x_1 + x_2}{2}, \frac{y_1 + y_2}{2}\right) = \left(\frac{-3 + 5}{2}, \frac{4 + 2}{2}\right)$$

$$= \left(\frac{2}{2}, \frac{6}{2}\right)$$

$$= (1, 3)$$

The midpoint of $\overline{AB}$ is $M(1, 3)$.

Practice

Simplify each expression.

1. $\dfrac{-1 + 3}{2}$ 2. $\dfrac{6 + 8}{7}$ 3. $\dfrac{12 + 4}{4}$

4. $\dfrac{5 + 4}{3}$ 5. $\dfrac{-5 + 8}{3}$ 6. $\dfrac{4 + 6}{5}$

Find the midpoint of $\overline{XY}$.

7. $X(3, 2)$ and $Y(-5, 0)$ 8. $X(6, 3)$ and $Y(4, 5)$ 9. $X(6, 5)$ and $Y(2, 3)$

10. $X(4, 5)$ and $Y(2, 3)$ 11. $X(3, 5)$ and $Y(-3, -5)$ 12. $X(7, 4)$ and $Y(3, 6)$

13. $X(5, 3)$ and $Y(-5, -3)$ 14. $X(0, 6)$ and $Y(0, 0)$ 15. $X(8, 0)$ and $Y(0, 10)$

16. $X(2, -3)$ and $Y(4, -3)$ 17. $X(4, 5)$ and $Y(-4, -5)$ 18. $X(-1, 0)$ and $Y(-7, 4)$

19. $X(1, 2)$ and $Y(1, 4)$ 20. $X(1, 6)$ and $Y(15, 8)$ 21. $X(2, 1)$ and $Y(2, 5)$

22. $X(3, 6)$ and $Y(-3, -6)$ 23. $X(0, 1)$ and $Y(0, 7)$ 24. $X(-1, 6)$ and $Y(1, -6)$

25. Pony Express stations were about 10 mi apart. The latitude-longitude coordinates of two former stations in western Nevada are (39.5N, 117.8W) and (39.3N, 118.2W). These two stations were about 20 mi apart. Estimate the coordinates of the station that was between them.

Lesson 9-3 Trigonometric Ratios

Finding Trigonometric Ratios

In $\triangle ABC$, $\overline{BC}$ is the side opposite $\angle A$ and $\overline{AC}$ is the side adjacent to $\angle A$. The hypotenuse is $\overline{AB}$. You can use these relationships to express trigonometric ratios.

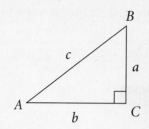

sine of $\angle A$ = $\dfrac{\text{length of side opposite } \angle A}{\text{length of hypotenuse}}$ or $\sin A = \dfrac{a}{c}$

cosine of $\angle A$ = $\dfrac{\text{length of side adjacent } \angle A}{\text{length of hypotenuse}}$ or $\cos A = \dfrac{b}{c}$

tangent of $\angle A$ = $\dfrac{\text{length of side opposite } \angle A}{\text{length of side adjacent to } \angle A}$ or $\tan A = \dfrac{a}{b}$

Example

Use the diagram below. Find $\sin B$, $\cos B$, $\tan B$.

$\sin B = \dfrac{6}{10} = \dfrac{3}{5}$

$\cos B = \dfrac{8}{10} = \dfrac{4}{5}$

$\tan B = \dfrac{6}{8} = \dfrac{3}{4}$

 Cross multiply to solve.

1. $\dfrac{1}{2} = \dfrac{4}{x}$ 2. $\dfrac{2}{5} = \dfrac{6}{x}$ 3. $\dfrac{4}{5} = \dfrac{8}{x}$

4. $\dfrac{3}{4} = \dfrac{6}{x}$ 5. $\dfrac{1}{3} = \dfrac{3}{x}$ 6. $\dfrac{5}{4} = \dfrac{10}{x}$

Use $\triangle XYZ$, $\triangle ABC$, or $\triangle RST$ to evaluate each expression.

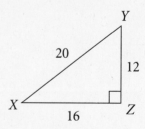 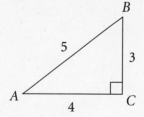 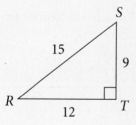

7. $\sin X$ 8. $\cos Y$ 9. $\tan S$

10. $\sin B$ 11. $\tan Y$ 12. $\sin S$

13. $\cos X$ 14. $\sin Y$ 15. $\cos S$

16. $\sin A$ 17. $\cos B$ 18. $\tan B$

19. $\cos A$ 20. $\cos R$ 21. $\tan R$

22. $\tan X$ 23. $\tan A$ 24. $\sin R$

Lesson 9-3 Trigonometric Ratios

Part
2

Solving Problems Using Trigonometric Ratios

You can use trigonometric ratios to measure distances indirectly when you know an angle of elevation.

An *angle of elevation* is an angle from the horizontal up to the line of sight.

Example

Suppose the angle of elevation from you to the top of a building is 75°. The building is 100 ft tall. How far from the building are you?

$$\tan 75° = \frac{\text{length of side opposite } \angle A}{\text{length of side adjacent to } \angle A}$$

$$\tan 75° = \frac{100}{x}$$

$$3.732050808 = \frac{100}{x} \quad \longleftarrow \textbf{Use a calculator.}$$

$$3.732050808x = 100 \quad \longleftarrow \textbf{Cross multiply.}$$

$$x = \frac{100}{3.732050808} \quad \longleftarrow \textbf{Divide.}$$

$$x = 26.79491924 \quad \longleftarrow \textbf{Use a calculator.}$$

$$x \approx 27$$

You are about 27 ft from the building.

Practice

Solve each equation for the variable *x*.

1. $\frac{10}{x} = 100$ **2.** $\frac{x}{20} = 0.34$ **3.** $0.82x = 40$

4. $\frac{x}{0.52} = 16$ **5.** $\frac{8}{7} = \frac{x}{35}$ **6.** $4x = 36$

Find the value of *x* to the nearest tenth.

7.

8.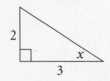

9.

10. A kite string that is 50 m long makes a 70° angle with the ground. How high above the ground is the kite?

11. Tye stands 12 ft from a 25 ft-tall tree. What is the angle of elevation from Tye's feet to the top of the tree?

12. An access ramp forms a 3° angle with the ground. The ramp rises 2 ft. How long is the ramp?

13. Marti walks 400 m east from her home to the bank. Then she walks 200 m north from the bank to a restaurant. How long is a straight path from the restaurant to Marti's house?

© Prentice-Hall, Inc.

Lesson 9-4 Simplifying Radicals

Multiplication with Radicals

Part **1**

Multiplication Property of Square Roots

For any numbers $a \geq 0$ and $b \geq 0$, $\sqrt{ab} = \sqrt{a} \cdot \sqrt{b}$.

Example

Simplify the radical expression $\sqrt{50}$.

$$\sqrt{50} = \sqrt{25 \cdot 2} \quad \longleftarrow \quad \textbf{25 is a perfect square and a factor of 50.}$$

$$= \sqrt{25} \cdot \sqrt{2} \quad \longleftarrow \quad \textbf{Use the multiplication property.}$$

$$= 5\sqrt{2} \quad \longleftarrow \quad \textbf{Simplify } \sqrt{25}.$$

Practice

Find XY.

1. $X(0, 0)$ and $Y(-3, 4)$ **2.** $X(5, 12)$ and $Y(0, 0)$ **3.** $X(2, 2)$ and $Y(5, -2)$

4. $X(1, 4)$ and $Y(6, 9)$ **5.** $X(3, 1)$ and $Y(-1, 6)$ **6.** $X(5, 6)$ and $Y(4, -1)$

Simplify each radical expression.

7. $\sqrt{32}$ **8.** $\sqrt{20}$ **9.** $\sqrt{27}$

10. $\sqrt{300}$ **11.** $\sqrt{18}$ **12.** $\sqrt{45}$

13. $\sqrt{8}$ **14.** $\sqrt{12}$ **15.** $\sqrt{28}$

16. $\sqrt{24}$ **17.** $\sqrt{48}$ **18.** $\sqrt{98}$

19. $\sqrt{54}$ **20.** $\sqrt{150}$ **21.** $\sqrt{1000}$

22. $\sqrt{200}$ **23.** $\sqrt{75}$ **24.** $\sqrt{63}$

25. $-\sqrt{12}$ **26.** $\sqrt{250}$ **27.** $\sqrt{72}$

28. $\sqrt{2} \cdot \sqrt{8}$ **29.** $\sqrt{12} \cdot \sqrt{3}$ **30.** $\sqrt{4} \cdot \sqrt{3}$

31. $3\sqrt{50} \cdot \sqrt{2}$ **32.** $2\sqrt{3} \cdot \sqrt{30}$ **33.** $-5\sqrt{6} \cdot 2\sqrt{3}$

34. $3\sqrt{3} \cdot 5\sqrt{15}$ **35.** $\sqrt{5} \cdot \sqrt{10}$ **36.** $\left(3\sqrt{2}\right)^2$

37. $\sqrt{3} \cdot \sqrt{12}$ **38.** $\sqrt{4 \cdot 2}$ **39.** $\sqrt{15} \cdot 2\sqrt{5}$

40. $\sqrt{9} \cdot \sqrt{4}$ **41.** $\sqrt{10} \cdot \sqrt{25}$ **42.** $\sqrt{16} \cdot 2\sqrt{3}$

43. $\sqrt{25} \cdot \sqrt{2}$ **44.** $\sqrt{6} \cdot \sqrt{3}$ **45.** $\sqrt{8} \cdot \sqrt{26}$

Find the area. Express in simplest form.

46. A triangle has a base of $2\sqrt{3}$ *m* and a height of $2\sqrt{2}$ *m*. Find the area.

47. The side of a square is $4\sqrt{10}$ ft. What is the area?

Lesson 9-4 Simplifying Radicals

Part
2 ▼

Division with Radicals

Division Property of Square Roots

For any numbers $a \geq 0$ and $b > 0$, $\sqrt{\frac{a}{b}} = \frac{\sqrt{a}}{\sqrt{b}}$

Sometimes it is easier to divide first, then simplify the radical expression.

Example

Simplify the radical expression $\sqrt{\frac{120}{15}}$.

$$\frac{\sqrt{120}}{\sqrt{15}} = \sqrt{\frac{120}{15}} \qquad \longleftarrow \textbf{Use the division property.}$$

$$= \sqrt{8} \qquad \longleftarrow \textbf{Divide.}$$

$$= \sqrt{4 \cdot 2} \qquad \longleftarrow \textbf{4 is a perfect square and a factor of 8.}$$

$$= \sqrt{4} \cdot \sqrt{2} \qquad \longleftarrow \textbf{Use the multiplication property.}$$

$$= 2\sqrt{2} \qquad \longleftarrow \textbf{Simplify } \sqrt{4} \text{ .}$$

Practice

Find each value.

1. $\sqrt{100}$

2. $-3\sqrt{9}$

3. $3\sqrt{64}$

4. $2\sqrt{121}$

5. $\sqrt{196}$

6. $2\sqrt{49}$

Simplify each radical expression.

7. $\sqrt{\frac{3}{4}}$

8. $\sqrt{\frac{7}{9}}$

9. $\sqrt{\frac{11}{16}}$

10. $\sqrt{\frac{49}{121}}$

11. $\sqrt{\frac{12}{81}}$

12. $\sqrt{\frac{35}{5}}$

13. $\sqrt{\frac{5}{9}}$

14. $\sqrt{\frac{20}{5}}$

15. $\sqrt{\frac{8}{36}}$

16. $\sqrt{\frac{3}{16}}$

17. $\sqrt{\frac{48}{4}}$

18. $\sqrt{\frac{36}{4}}$

19. $\sqrt{\frac{52}{4}}$

20. $\sqrt{\frac{20}{81}}$

21. $\sqrt{\frac{10}{25}}$

22. $\frac{\sqrt{27}}{\sqrt{3}}$

23. $\frac{\sqrt{90}}{\sqrt{10}}$

24. $\frac{\sqrt{75}}{\sqrt{3}}$

25. $\frac{\sqrt{24}}{\sqrt{6}}$

26. $\frac{\sqrt{2}}{\sqrt{2}}$

27. $\sqrt{\frac{3}{49}}$

28. $\frac{\sqrt{40}}{\sqrt{10}}$

29. $\sqrt{\frac{4}{25}}$

30. $\sqrt{\frac{7}{36}}$

31. $\sqrt{\frac{10}{81}}$

32. $\sqrt{\frac{6}{16}}$

33. $\sqrt{\frac{7}{25}}$

34. $\sqrt{\frac{6}{100}}$

35. $\sqrt{\frac{11}{144}}$

36. $\sqrt{\frac{21}{27}}$

© Prentice-Hall, Inc.

Lesson 9-5 Adding and Subtracting Radicals
Simplifying Sums and Differences

Part
1

You can simplify radical expressions by combining like terms. Like terms have the same radical part.

like terms	unlike terms
$3\sqrt{5}$ and $-4\sqrt{5}$	$4\sqrt{13}$ and $2\sqrt{3}$

Example

Simplify the expression $7\sqrt{2} - 2\sqrt{8}$.

$7\sqrt{2} - 2\sqrt{8} = 7\sqrt{2} - 2\sqrt{4 \cdot 2}$ ← **4 is a perfect square and a factor of 8.**

$\qquad\qquad\quad = 7\sqrt{2} - 2\sqrt{4} \cdot \sqrt{2}$ ← **Use the multiplication property.**

$\qquad\qquad\quad = 7\sqrt{2} - 4\sqrt{2}$ ← **Simplify $\sqrt{4}$.**

$\qquad\qquad\quad = 5\sqrt{2}$ ← **Combine like terms.**

Practice

Simplify each expression.

1. $2x - 3 + 5x$ **2.** $-4x - 5 - 7x$ **3.** $3 - 3y + 4$

4. $3y + 2 + y$ **5.** $10y - 8y - 2$ **6.** $8y - 4 + 3$

Simplify each expression.

7. $5\sqrt{3} - \sqrt{12}$ **8.** $\sqrt{18} - 6\sqrt{2}$ **9.** $6\sqrt{2} - 4\sqrt{2}$

10. $2\sqrt{11} + 5\sqrt{11}$ **11.** $5\sqrt{2} - 4\sqrt{8}$ **12.** $5\sqrt{7} + 3\sqrt{7}$

13. $4\sqrt{3} - \sqrt{27}$ **14.** $7\sqrt{3} - \sqrt{27}$ **15.** $3\sqrt{4} - 6$

16. $3\sqrt{3} + 2\sqrt{12}$ **17.** $3\sqrt{7} - \sqrt{7}$ **18.** $\sqrt{20} - \sqrt{5}$

19. $4\sqrt{7} - 5\sqrt{7}$ **20.** $3\sqrt{2} - \sqrt{18}$ **21.** $\sqrt{36} - \sqrt{49}$

22. $\sqrt{5} + \sqrt{5}$ **23.** $2\sqrt{8} + 2\sqrt{2}$ **24.** $\sqrt{2} + 3\sqrt{2}$

25. $2\sqrt{5} - \sqrt{20}$ **26.** $4\sqrt{5} + 6\sqrt{5}$ **27.** $\sqrt{90} - \sqrt{40}$

28. $\sqrt{24} + 5\sqrt{6}$ **29.** $11\sqrt{10} - 10\sqrt{10}$ **30.** $\sqrt{5} + 2\sqrt{20}$

31. $5\sqrt{2} + \sqrt{2}$ **32.** $\sqrt{3} - 4\sqrt{3}$ **33.** $\sqrt{6} - 4\sqrt{6}$

34. $-2\sqrt{2} - \sqrt{2}$ **35.** $8\sqrt{27} - 3\sqrt{3}$ **36.** $9\sqrt{50} - 4\sqrt{2}$

37. $5\sqrt{50} - 2\sqrt{18}$ **38.** $3\sqrt{18} + 9$ **39.** $-\sqrt{3} - 2\sqrt{3}$

40. $\sqrt{12} + 3\sqrt{3}$ **41.** $\sqrt{5} - 6\sqrt{5}$ **42.** $7\sqrt{5} - \sqrt{20}$

43. $\sqrt{8} - \sqrt{2}$ **44.** $\sqrt{50} - 4\sqrt{2}$ **45.** $\sqrt{5} - \sqrt{20}$

Lesson 9-5 Adding and Subtracting Radicals

Part **2**

Simplifying Products, Sums, and Differences

Sometimes you need to use the distributive property to simplify expressions.

Example

Simplify $\sqrt{3}\,(8 - \sqrt{12}\,)$.

$$\sqrt{3}\,(8 - \sqrt{12}\,) = \sqrt{3}\,(8) - \sqrt{3}\,(\sqrt{12}\,) \quad \longleftarrow \text{ Use the distributive property.}$$
$$= 8\sqrt{3} - \sqrt{3 \cdot 12} \quad \longleftarrow \text{ Use the multiplication property.}$$
$$= 8\sqrt{3} - \sqrt{36} \quad \longleftarrow \text{ Multiply.}$$
$$= 8\sqrt{3} - 6 \quad \longleftarrow \text{ Simplify.}$$

Practice

Simplify each expression.

1. $x(4 - y)$ **2.** $-6(-4 - y)$ **3.** $y(9 - x)$

4. $3(x + 2)$ **5.** $5(x + 3)$ **6.** $x(5 + y)$

Simplify each expression.

7. $\sqrt{3}\,(3\sqrt{2} + \sqrt{3}\,)$ **8.** $\sqrt{3}\,(5\sqrt{3} - 2\sqrt{6}\,)$ **9.** $\sqrt{2}\,(2\sqrt{2} - 6)$

10. $\sqrt{2}\,(5 + 3\sqrt{6}\,)$ **11.** $\sqrt{6}\,(5\sqrt{2} + 6)$ **12.** $-3\sqrt{2}\,(2\sqrt{2} - \sqrt{6}\,)$

13. $5\sqrt{2}\,(3\sqrt{8} - 3\sqrt{2}\,)$ **14.** $\sqrt{5}\,(3 - 2\sqrt{5}\,)$ **15.** $-\sqrt{3}\,(5 + 2\sqrt{3}\,)$

16. $\sqrt{2}\,(2\sqrt{6} - 3)$ **17.** $\sqrt{3}\,(4 + \sqrt{3}\,)$ **18.** $\sqrt{6}\,(2\sqrt{3} - 4\sqrt{2}\,)$

19. $\sqrt{7}\,(5 + 3\sqrt{5}\,)$ **20.** $\sqrt{5}\,(\sqrt{5} - 3)$ **21.** $\sqrt{6}\,(\sqrt{6} - 2)$

22. $\sqrt{2}\,(2 + \sqrt{2}\,)$ **23.** $\sqrt{2}\,(2\sqrt{3} - 3\sqrt{4}\,)$ **24.** $\sqrt{2}\,(2 - 3\sqrt{6}\,)$

25. $3\sqrt{2}\,(\sqrt{2} + 2\sqrt{8}\,)$ **26.** $2\sqrt{3}\,(\sqrt{2} - \sqrt{5}\,)$ **27.** $-3\sqrt{3}\,(\sqrt{2} - \sqrt{3}\,)$

28. $\sqrt{2}\,(\sqrt{3} + \sqrt{5}\,)$ **29.** $\sqrt{3}\,(2\sqrt{3} + \sqrt{27}\,)$ **30.** $4\sqrt{5}\,(2 + 3\sqrt{5}\,)$

31. $-5\sqrt{3}\,(-3\sqrt{5} + \sqrt{3}\,)$ **32.** $4\sqrt{3}\,(\sqrt{2} + \sqrt{3}\,)$ **33.** $\sqrt{5}\,(3 + \sqrt{10}\,)$

34. $\sqrt{2}\,(\sqrt{3} + 3)$ **35.** $\sqrt{7}\,(\sqrt{2} + 5)$ **36.** $\sqrt{4}\,(\sqrt{8} + 2)$

37. $\sqrt{3}\,(5 + \sqrt{6}\,)$ **38.** $\sqrt{6}\,(3 + \sqrt{2}\,)$ **39.** $\sqrt{3}\,(\sqrt{6} + \sqrt{5}\,)$

40. $\sqrt{2}\,(5 - \sqrt{2}\,)$ **41.** $\sqrt{7}\,(7 - \sqrt{7}\,)$ **42.** $\sqrt{3}\,(\sqrt{2} - \sqrt{6}\,)$

43. $4(\sqrt{3} - \sqrt{9}\,)$ **44.** $2(\sqrt{3} - \sqrt{12}\,)$ **45.** $10(\sqrt{3} - 4)$

46. $1(\sqrt{3} + 2)$ **47.** $\sqrt{3}\,(\sqrt{2} + \sqrt{3}\,)$ **48.** $\sqrt{7}\,(8 - \sqrt{9}\,)$

49. $\sqrt{3}\,(3 - \sqrt{9}\,)$ **50.** $\sqrt{5}\,(5 - \sqrt{25}\,)$ **51.** $\sqrt{5}\,(4 - \sqrt{2}\,)$

Lesson 9-6 Solving Radical Equations

Part 1

Solving a Radical Equation

An equation that has a variable under a radical is a *radical equation*. You can often solve a radical equation by squaring both sides. To do this, first get the radical by itself on one side of the equation. Remember that the expression under the radical must be positive.

$$\text{when } x \geq 0, (\sqrt{x})^2 = x$$

Example

Solve $\sqrt{x} + 8 = 17$. Check your solution.

$$\sqrt{x} + 8 = 17$$

$$\sqrt{x} + 8 - 8 = 17 - 8 \qquad \longleftarrow \textbf{Subtract 8 from both sides.}$$

$$\sqrt{x} = 9 \qquad \longleftarrow \textbf{Simplify.}$$

$$(\sqrt{x})^2 = (9)^2 \qquad \longleftarrow \textbf{Square both sides.}$$

$$x = 81$$

Check. $\sqrt{x} + 8 = 17$

$$\sqrt{81} + 8 \overset{?}{=} 17 \qquad \longleftarrow \textbf{Substitute 81 for } x.$$

$$9 + 8 = 17 \checkmark$$

The solution of $\sqrt{x} + 8 = 17$ is 81.

Practice

Evaluate each expression for the given value.

1. $\sqrt{x} + 2$ for $x = 121$ **2.** $\sqrt{x + 6}$ for $x = 3$ **3.** $2\sqrt{x + 1}$ for $x = 8$

4. $2\sqrt{x}$ for $x = 9$ **5.** $\sqrt{x} + 3$ for $x = 16$ **6.** $\sqrt{x + 7}$ for $x = 1$

Solve each radical equation. Check your solution.

7. $\sqrt{x} = 3\sqrt{2}$ **8.** $\sqrt{x} = 2\sqrt{5}$ **9.** $\sqrt{x - 2} = 3$

10. $\sqrt{x + 4} = 6$ **11.** $\sqrt{4x} = 4$ **12.** $3\sqrt{x} = 6$

13. $\sqrt{x} - 8 = 1$ **14.** $\sqrt{2x} = 4$ **15.** $\sqrt{5x} = 5$

16. $\sqrt{4x} = 8$ **17.** $\sqrt{3x - 2} = 4$ **18.** $8 - \sqrt{x} = 1$

19. $\sqrt{5x + 6} = 1$ **20.** $\sqrt{10 + x} = 5$ **21.** $\sqrt{3 - x} = 2$

22. $\sqrt{x} + 3 = 6$ **23.** $\sqrt{x} = 36$ **24.** $6 - \sqrt{x - 5} = 3$

25. $\sqrt{x - 1} = 2$ **26.** $\sqrt{5x + 4} = 3$ **27.** $-\sqrt{x} = -5$

28. $-9 = -\sqrt{x}$ **29.** $\sqrt{x} = \sqrt{5}$ **30.** $\sqrt{x - 1} = \sqrt{3}$

31. $\sqrt{x} + 4 = -1$ **32.** $16 - \sqrt{x} = 10$ **33.** $9 - \sqrt{x} = 8$

Lesson 9-6 Solving Radical Equations

Part 2

Solving Equations with Extraneous Solutions

When you solve equations by squaring both sides, you sometimes find two possible solutions. You need to determine which solution actually satisfies the original equation.

Example

Solve $x = \sqrt{x + 2}$.

$$x = \sqrt{x + 2}$$
$$(x)^2 = (\sqrt{x + 2})^2 \qquad \longleftarrow \textbf{Square both sides.}$$
$$x^2 = x + 2$$
$$x^2 - x - 2 = 0 \qquad \longleftarrow \textbf{Subtract } x \textbf{ and 2 from both sides.}$$
$$x = \frac{-(-1) \pm \sqrt{(-1)^2 - 4(1)(-2)}}{2(1)} \qquad \longleftarrow \textbf{Use the quadratic formula to solve for } x.$$
$$x = \frac{1 \pm \sqrt{1 - (-8)}}{2}$$
$$x = \frac{1 \pm \sqrt{9}}{2}$$
$$x = \frac{1 \pm 3}{2} \qquad \longleftarrow \textbf{Simplify } \sqrt{9}.$$
$$x = \frac{1 + 3}{2} \text{ or } \frac{1 - 3}{2}$$
$$x = 2 \text{ or } x = -1 \qquad \longleftarrow \textbf{Two possible solutions}$$

Check.
$$x = \sqrt{x + 2}$$
$$2 \overset{?}{=} \sqrt{2 + 2} \qquad -1 \overset{?}{=} \sqrt{-1 + 2}$$
$$2 = 2 ✔ \qquad\qquad -1 \neq 1$$

The correct solution is 2.

Practice

Simplify.

1. $5\sqrt{2} + \sqrt{2}$ **2.** $\sqrt{3} - 4\sqrt{3}$ **3.** $\sqrt{x} - 6\sqrt{x}$

4. $4\sqrt{3} + \sqrt{3}$ **5.** $5\sqrt{3} - 2\sqrt{3}$ **6.** $-5\sqrt{b} + \sqrt{4b}$

Solve each radical equation. Check your solutions.

7. $\sqrt{4x^2 - 27} = x$ **8.** $5\sqrt{x} = -5$ **9.** $\sqrt{4x^2 - 15} = x$

10. $6 + \sqrt{x - 5} = 3$ **11.** $-\sqrt{x} = 6$ **12.** $\sqrt{x + 5} = 4$

13. $3 + \sqrt{x - 1} = 5$ **14.** $\sqrt{x^2 + 5} = x + 1$ **15.** $\sqrt{2x + 3} = \sqrt{x}$

16. $\sqrt{x - 3} = x$ **17.** $\sqrt{2x + 3} = x$ **18.** $\sqrt{x - 5} = 2$

19. $\sqrt{12 - 4x} = x$ **20.** $x = \sqrt{2x + 8}$ **21.** $x = \sqrt{6x - 8}$

22. $x = \sqrt{5 - 4x}$ **23.** $x = \sqrt{x + 12}$ **24.** $x = \sqrt{6 - x}$

© Prentice-Hall, Inc.

Lesson 9-7 Graphing Square Root Functions

Example

Find the domain of $y = \sqrt{x - 2}$. Then graph the function.

The square root limits the domain because the expression under the radical cannot be negative. To find the domain, solve $x - 2 \geq 0$.

$$x - 2 \geq 0$$
$$x \geq 2$$

The domain is the set of all real numbers greater than or equal to 2. Use a table of values or graphing calculator to graph.

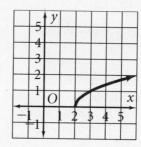

x	$\sqrt{x - 2}$
2	$\sqrt{2 - 2} = 0$
3	$\sqrt{3 - 2} = 1$
4	$\sqrt{4 - 2} = 1.41$
5	$\sqrt{5 - 2} = 1.73$
6	$\sqrt{6 - 2} = 2$

Practice

Find the value of y for the given value of x.

1. $y = \sqrt{x + 7} - 3$ for $x = 2$ 2. $y = 3\sqrt{x} + 3$ for $x = 16$

3. $y = \sqrt{x} - 1$ for $x = 9$ 4. $y = 2\sqrt{x}$ for $x = 9$

5. $y = \sqrt{x + 2}$ for $x = 14$ 6. $y = \sqrt{x + 4}$ for $x = 21$

Find the domain of each function. Then graph the function.

7. $y = \sqrt{x} + 1$ 8. $y = \sqrt{x + 2}$ 9. $y = \sqrt{x - 6}$

10. $y = \sqrt{x + 2} - 1$ 11. $y = \sqrt{x + 3} - 1$ 12. $y = \sqrt{x} - 4$

13. $y = \sqrt{x - 7} + 6$ 14. $y = \sqrt{x - 5} + 4$ 15. $y = \sqrt{x + 1} - 8$

16. $y = \sqrt{x + 4} + 3$ 17. $y = 3\sqrt{x}$ 18. $y = \sqrt{x - 1} - 1$

19. $y = \sqrt{x - 3}$ 20. $y = \sqrt{x + 5}$ 21. $y = \sqrt{x}$

22. $y = \sqrt{x + 6} - 5$ 23. $y = \sqrt{x - 2} - 2$ 24. $y = \sqrt{x - 3} - 2$

25. $y = \sqrt{x - 4} - 3$ 26. $y = \sqrt{x - 5} - 3$ 27. $y = \sqrt{x + 1}$

28. $y = \sqrt{x + 7} - 4$ 29. $y = \sqrt{x} - 5$ 30. $y = \sqrt{x} + 4$

31. $y = 2\sqrt{x} - 3$ 32. $y = \sqrt{x - 6} - 2$ 33. $y = 4\sqrt{x}$

34. $y = \sqrt{x - 3} + 3$ 35. $y = \sqrt{x - 3} + 1$ 36. $y = \sqrt{x - 4} - 4$

37. $y = \sqrt{x - 4} + 2$ 38. $y = \sqrt{x} + 2$ 39. $y = \sqrt{x - 2} + 3$

40. $y = \sqrt{x} - 3$ 41. $y = \sqrt{x + 3} + 3$ 42. $y = \sqrt{x} + 3$

43. $y = \sqrt{x} - 2$ 44. $y = \sqrt{x} - 1$ 45. $y = \sqrt{x} - 4$

Name _____ Class _____ Date _____

Lesson 9-8 Analyzing Data Using Standard Deviation

You can calculate the standard deviation by following these steps.

Step 1 Find the mean of the data. The expression $\bar{x}$ represents the mean.

Step 2 Find the difference of each data value from the mean.

Step 3 Calculate the square of each difference.

Step 4 Find the sum of the squares of the differences.

Step 5 Divide the sum by the number of data values.

Step 6 Take the square root of the quotient just calculated.

Example

Find the standard deviation of the data set {94, 86, 90, 88, 76}.

Calculate the standard deviation.

x	$\bar{x}$	$x - \bar{x}$	$(x - \bar{x})^2$
94	86.8	7.2	51.84
86	86.8	−0.8	0.64
90	86.8	3.2	10.24
88	86.8	1.2	1.44
76	86.8	−10.8	116.64
		Sum	180.8

sum of squares $= 180.8$

$\dfrac{\text{sum of squares}}{5} = \dfrac{180.8}{5}$

$= 36.16$

$\sqrt{36.16} = 6.0133186$

The standard deviation is about 6.0 for the data set.

Practice

Find the mean to the nearest hundredth.

1. {1, 2, 4}　　　　　**2.** {10, 12, 14}　　　　**3.** {6, 4, 5}

4. {85, 80, 90}　　　**5.** {2, 1, 0}　　　　　　**6.** {50, 70, 60}

Make a table to find the standard deviation for each data set.

7. {6, 3, 4, 5, 2}　　　　　**8.** {−5.1, −3.6, 0, −2.5, −1}　　**9.** {17, 19, 24, 13, 25}

10. {10, 8.5, 7, 9, 11}　　　**11.** {−6.5, −3, 4, 0, 3}　　　　**12.** {45, 49, 38, 46, 44}

13. {95, 111, 98.9, 97, 88}　**14.** {−30, −25, −22, −32}　　**15.** {0.2, 0.4, 0.9, 2, −0.1}

16. {12, 14, 13, 12.5, 13.1}　**17.** {7, 27, 17, 37, 47}　　　**18.** {99, 120, 86, 80, 100}

19. {10, 20, 30, 40, 50}　　　**20.** {1.1, 1.2, 1.3, 1.4}　　　**21.** {101, 107, 105, 100}

22. {5, 15, 10, 20}　　　　　**23.** {12, 11, 13, 10}　　　　　**24.** {36, 25, 14}

Lesson 10-1 Adding and Subtracting Polynomials

Part **1**

Describing Polynomials

A **polynomial** is one term or the sum or difference of two or more terms. A polynomial has no variables in the denominator. For a term that has only one variable, the **degree of a term** is the exponent of the variable.

$$2x + x^3 - 8x^2 - 5$$

degree ⟶   1   3   2   0   ⟵ The degree of a constant is 0.

The **degree of a polynomial** is the same as the term with the highest degree. You can name a polynomial by its degree or by the number of its terms.

The polynomial is in **standard form** when the terms decrease in degree from left to right and no terms have the same degree.

Example

Write each polynomial in standard form. Then name each polynomial by its degree and the number of its terms.

a. $7 + 4x$

$4x + 7$

linear binomial

b. $5x^3 - 3 + x^2$

$5x^3 + x^2 - 3$

cubic trinomial

c. $4x + 7 - 2x^4$

$-2x^4 + 4x + 7$

fourth degree trinomial

Practice

 Mini Help

Arrange in order from greatest to least.

1. $-4, 2, -3, 5$

2. $3, -7, 1, 9$

3. $-1, -5, 0, -3$

4. $14, -14, 12, -12$

5. $-3, -5, 2, 8$

6. $7, -8, 10, -9$

Write each polynomial in standard form. Then name each polynomial by its degree and number of terms.

7. $7x + 4x^2$

8. $9x - 5$

9. $10 - 12x + 5x^2$

10. $8y^2 - 3y^3 + y - 1$

11. $3k^4 + k^6 - 2$

12. $-2a + 7a^7 - 3a^3 - 12$

13. $b^2 + 5b - b^3$

14. $2 + 11c^3$

15. $3g^2 - 4 - 5g$

16. $17c + 13c^5 + 12c^2$

17. $6x^2 - 3 - x$

18. $-9 + 5y^5 + 2y^2 - y$

19. $14 - 12x^3 + 5x$

20. x^3

21. $y + 4$

22. $y^5 - 5$

23. $-x^2 + 5 - 6x + 4x^3$

24. $4p^2 + 8 - 5p$

25. $-6 + 2p^2$

26. y^4

27. $x^4 + x^{10} + x^{14}$

28. $95 + b^3$

29. $-3 + x^4 + x^2$

30. $g + g^3 - 3 + 4g^2$

31. $25 - 4x^5$

32. x

33. $5a^2 - 8a^{10} - a^5 + a^7 - 7$

Lesson 10-1 Adding and Subtracting Polynomials

Part 2

Adding Polynomials

There are three methods that can be used to add polynomials. You can use tiles, add vertically, or add horizontally.

Opposite terms form zero pairs ⟶ ☐ ■ = 0   ☐ ▮ = 0   ☐ ■ = 0

x^2 $-x^2$ x $-x$ 1 -1

Example

Find $(3x^2 + 2x + 1) + (x^2 - 4x - 3)$.

Method 1 Add using tiles.

$3x^2 + 2x + 1$

$x^2 - 4x - 3$

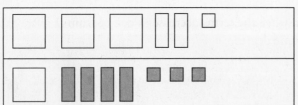

Group like tiles together. Remove zero pairs. Then count the tiles that are left.

$4x^2 - 2x - 2$

Method 2 Add vertically.
Line up like terms. Then add the coefficients.

$$3x^2 + 2x + 1$$
$$\underline{1x^2 - 4x - 3}$$
$$4x^2 - 2x - 2$$

Method 3 Add horizontally.
Group the like terms. Then add the coefficients.

$$(3x^2 + 2x + 1) + (x^2 - 4x - 3)$$
$$= (3x^2 + x^2) + (+2x - 4x) + (1 - 3)$$
$$= 4x^2 - 2x - 2$$

Practice

Combine like terms and simplify.

1. $x + y + 2x - 3y$ **2.** $-c + 3d + 2c - 4d$ **3.** $7x + 5 - 4x - 3$

4. $8d - 7c + 2 - d$ **5.** $10a - 5 + 5a + 4$ **6.** $-5 - 3t + 2u - 4t$

Find each sum using one of the three methods above.

7. $(4x^2 + 7x - 3) + (x^2 - 4x + 3)$ **8.** $(5x^2 - 12x + 10) + (-4x^2 + 8x - 7)$

9. $(8y^2 - 3y - 1) + (-6y^2 - y)$ **10.** $(3a^3 + a^2 - 2a - 5) + (a^3 + 3a^2 - 7a + 4)$

11. $(b^2 + 5b - 3) + (3b^2 - 5b + 4)$ **12.** $(2x^2 + 3x + 7) + (3x^2 - x - 4)$

13. $(2x + 5) + (-4x - 5)$ **14.** $(y^2 - 5y - 2) + (y^2 + y - 2)$

Lesson 10-1 Adding and Subtracting Polynomials

Subtracting Polynomials

Part 3

When you subtract a polynomial, change each term of the polynomial being subtracted to its opposite. Then add the coefficients.

Example

Find $(5x^4 + 3x^2 + 4x - 2) - (3x^4 - x^2 - 3)$.

Method 1 Subtract vertically.

Line up like terms.

$$5x^4 + 3x^2 + 4x - 2$$
$$\underline{-(3x^4 - x^2 \qquad - 3)}$$

$\longrightarrow$

Add the opposite.

$$5x^4 + 3x^2 + 4x \quad - 2$$
$$\underline{-3x^4 + x^2 \qquad + 3}$$
$$2x^4 + 4x^2 + 4x + 1$$

Method 2 Subtract horizontally.

Change each term in the polynomial being subtracted to its opposite. Group like terms. Then add the coefficients of like terms.

$$(5x^4 + 3x^2 + 4x - 2) - (3x^4 - x^2 - 3)$$
$$= 5x^4 + 3x^2 + 4x - 2 - 3x^4 + x^2 + 3$$
$$= (5x^4 - 3x^4) + (3x^2 + x^2) + 4x + (-2 + 3)$$
$$= 2x^4 + 4x^2 + 4x + 1$$

Practice

 Mini Help

Write the opposite of each term in each expression.

1. $-(-4)$

2. $-12 + a$

3. $-(-x + y) - z$

4. $-x - (-3)$

5. $4 - b - c$

6. $-5 - 2m$

Find each difference.

7. $(3x^3 + 5x - 2) - (x^3 - 4x - 3)$

8. $(2x^4 - 3x^2 + 12x - 10) - (-4x^4 + 8x - 7)$

9. $(y^2 - 7y) - (-4y^2 + 3y - 1)$

10. $(2a^3 + 9a - 2) - (a^2 + 4a - 7)$

11. $(b^4 - 5b^3 - 4b^2) - (b^4 - 5b^3 + 3)$

12. $(4x^3 - 3x + 8) - (x^3 + x^2 - 4)$

13. $(3y^3 - 3y + 5) - (y^3 - 2y - 9)$

14. $(x^2 + 2x + 5) - (2x^2 - 4x - 5)$

© Prentice-Hall, Inc.

Lesson 10-2
Multiplying and Factoring with Monomials

Part 1

Multiplying by a Monomial

You can use the distributive property to multiply polynomials. You can also use tiles to multiply polynomials.

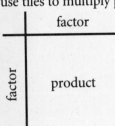

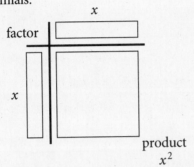

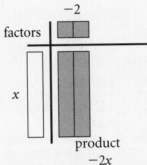

Example

Multiply $-2x$ and $(x - 1)$.

Method 1 Use tiles.

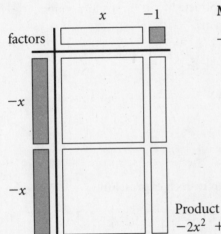

Method 2 Use the distributive property.

$$-2x(x - 1) = -2x(x) - 2x(-1)$$
$$= -2x^2 + 2x$$

Product
$-2x^2 + 2x$

Practice

Use the distributive property to simplify each expression.

1. $3(x + 2)$ 2. $5(a - 2)$ 3. $2(x + y)$

4. $(m - 7)2$ 5. $-(b - c)$ 6. $(x - 4)5$

Find each product.

7. $2x(5x - 3)$ 8. $(2x^2)(x^2 + 5x - 3)$ 9. $(y^2)(-4y - 1)$

10. $(3a^2)(a^2 + 4a - 7)$ 11. $(-4b^2)(b^4 - 5b^3 + 2)$ 12. $(-3x)(x^3 + x^2 - 4)$

13. $(3y)(y^2 + 2y - 7)$ 14. $(x^2)(2x^2 - 4x - 5)$ 15. $4x(x - 4)$

16. $x(-2x + 1)$ 17. $(-y^2)(y - 3)$ 18. $2b(2b - 2)$

19. $(-c^3)(c + 4)$ 20. $(x^4)(x - 1)$ 21. $(a^2)(a^2 + 1)$

22. $n(n^2 - 4)$ 23. $y(y - 1)$ 24. $5x(-x - 2)$

Lesson 10-2 Multiplying and Factoring with Monomials _{Part}

Factoring Out a Monomial

2

To factor out a monomial using the distributive property, you need to find
the greatest common factor (GCF).

Example

Factor $6x^4 - 30x^3 + 24x$.

Step 1 Find the GCF.
List the factors of each term.
Identify the factors common to all terms.

$$6x^4 = 2 \cdot 3 \cdot x \cdot x \cdot x \cdot x$$
$$30x^3 = 2 \cdot 3 \cdot 5 \cdot x \cdot x \cdot x$$
$$24x = 2 \cdot 2 \cdot 2 \cdot 3 \cdot x$$

The GCF is $2 \cdot 3 \cdot x$ or $6x$.

Step 2 Factor out the GCF.

$$6x^4 - 30x^3 + 24x$$
$$= 6x(x^3) - 6x(5x^2) + 6x(4)$$
$$= 6x(x^3 - 5x^2 + 4)$$

Practice

Find the GCF for each set of numbers.

1. 10, 25, 15   **2.** 14, 7, 28   **3.** $-4, -6, -8$

4. 16, 8, 12   **5.** $-9, -15, -12$ **6.** 12, 18, 24

Factor each polynomial.

7. $10x^2 - 25x$ **8.** $9x + 12$ **9.** $6y^3 - 15y^2 + 3y$

10. $14a^3 + 35a^2 + 21a$ **11.** $-b^4 + 5b^3 - 4b^2$ **12.** $51x^3 + 17x^2 - 34x$

13. $3y^3 + 2y^2 - 7y$ **14.** $12x^5 + 4x^4 - 36x^3 + 24x^2$ **15.** $6b + 18$

16. $7y + 21$ **17.** $4x^2 + x$ **18.** $8y^3 + 24y^2$

19. $7a^2 + 49a$ **20.** $5w^4 - 20w^3$ **21.** $4p^3 - 12p^2 + 18p$

22. $15t^8 - 45t^3$ **23.** $14a^4 - 42a^3 + 28a^2$ **24.** $12x^3 - 18x^2 + 24x$

25. $16x^2 - 12x + 24$ **26.** $3y^4 - 4y^3 + 2y^2$ **27.** $4x^2 + 12x + 8$

28. $a^2 + 7a$ **29.** $2p^3 + 4p^2 + 2p$ **30.** $8x - 8y$

31. $3k^2 - 9$ **32.** $8n^2 - 10n$ **33.** $4x^2 + 8x + 12$

34. $3x^3 + 6x^2 - 9x$ **35.** $3p + 6$ **36.** $10z + 15$

37. $8k - 4$ **38.** $27r + 9$ **39.** $7x^2 + 2x$

40. $14y - y^4$ **41.** $12 - 12a$ **42.** $2x^3 - 12x^2$

43. $5a^2 - 15a^3$ **44.** $7c + 14c^2$ **45.** $6x^4 + 18x^2$

46. $13x^3 + 26x^2$ **47.** $33w^3 + 11w^2$ **48.** $7a^2 - 9a^4$

49. $4j^3 - 6j^2 + 8j$ **50.** $5n - 15n^2 + 10$ **51.** $4a^2 + 8a^3 + 12a$

52. $7x^3 + 6x^2 + 5x$ **53.** $12m^2 + 27m^3 - 3m - 6$ **54.** $8r^5 - 4r^4 + 32r^3 + 6r$

55. $-3a^2 + 15a^3 - 18a$ **56.** $14x - 21x^4 + 35x^2 - 28$ **57.** $3a^2 - 3a$

Lesson 10-3 Multiplying Polynomials

Part 1

Multiplying Two Binomials

You can use tiles or the distributive property to multiply two binomials.

Example

Find the product $(x - 2)(3x - 2)$.

Method 1 Use tiles.

Step 1 Show the factors.

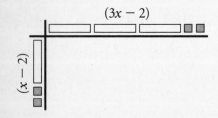

$(3x - 2)$

$(x - 2)$

Step 2 Find the product.

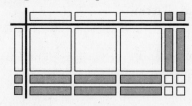

$3x^2 - 6x - 2x + 4$

$3x^2 - 8x + 4$ ← **Add coefficients of like terms.**

Method 2 Use the distributive property.

$(x - 2)(3x - 2) = x(3x) + x(-2) - 2(3x) - 2(-2)$

$= 3x^2 - 2x - 6x + 4$

$= 3x^2 - 8x + 4$

The product is $3x^2 - 8x + 4$.

Practice

Simplify.

1. $(2x)(x) - (2x)(3)$
2. $(4)(y) + (7)(y)$
3. $(-3)(a^2) + (-3)(2a)$

4. $(b)(3b) + (b)(-1)$
5. $(4y)(x) - (y)(x)$
6. $(-x)(7x) - (-x)(9)$

Find each product.

7. $(3a + 2)(a - 5)$
8. $(x - 4)(x - 7)$
9. $(6y + 5)(y + 1)$

10. $(2a - 5)(a + 3)$
11. $(2c - b)(c + 2b)$
12. $(8x + 5)(2x - 1)$

13. $(10y + 7)(3 - 2y)$
14. $(-5 - 2x)(7 - 3x)$
15. $(3p - 2)(p - 1)$

16. $(x + 2)(x - 1)$
17. $(-y - 2)(-y - 3)$
18. $(2 + x)(7 - x)$

19. $(9 - y)(y - 9)$
20. $(k + 2)(k + 5)$
21. $(x - 3)(x + 4)$

22. $(3y - 5)(2y - 1)$
23. $(10 + x)(11 + x)$
24. $(4y + 7)(y - 3)$

Lesson 10-3 Multiplying Polynomials

Part **2**

Multiplying Using FOIL

The term FOIL is a memory device for applying the distributive property
when multiplying two binomials. FOIL stands for "First, Outer, Inner, Last."

Example

Find the product $(4x - 9)(3x - 4)$.

$$
\begin{array}{cccccc}
& \textbf{FIRST} & \textbf{OUTER} & \textbf{INNER} & \textbf{LAST} \\
(4x - 9)(3x - 4) = & (4x)(3x) + & (4x)(-4) - & (9)(3x) - & (9)(-4) \\
= & 12x^2 & - \quad 16x & - \quad 27x & + \quad 36 \\
= & 12x^2 & - & 43x & + \quad 36
\end{array}
$$

The product is $12x^2 - 43x + 36$.   MIDDLE TERM

Practice

Simplify.

1. $(x)(x) - (x)(4)$ **2.** $(2)(y) + (3)(y)$ **3.** $(-1)(a) + (-1)(2a)$

4. $(3)(2b) + (b)(-1)$ **5.** $(2y)(2) - (y)(3)$ **6.** $(-2x)(2x) - (-x)(3)$

Find each product using FOIL.

7. $(y + 1)(y - 4)$ **8.** $(x + 5)(x + 3)$ **9.** $(x - 2)(x - 1)$

10. $(2a - 1)(a - 2)$ **11.** $(c + 3)(c + 4)$ **12.** $(x - 1)(2x - 10)$

13. $(y + 1)(3 - y)$ **14.** $(5 + x)(2 + 2x)$ **15.** $(6x + 1)(x - 4)$

16. $(4x - 8)(x + 2)$ **17.** $(x - 1)(x - 2)$ **18.** $(x + 3)(x + 4)$

19. $(x + 5)(x - 6)$ **20.** $(4x - 1)(x + 1)$ **21.** $(5x - 2)(x + 2)$

22. $(3y - 1)(y - 1)$ **23.** $(y + 10)(y - 9)$ **24.** $(4x - 5)(x + 2)$

25. $(-1 - x)(-2 + x)$ **26.** $(4 - y)(5 + y)$ **27.** $(5 + 2y)(y + 1)$

28. $(3x - 1)(x + 3)$ **29.** $(9x + 1)(x - 1)$ **30.** $(4 + x)(10 + x)$

31. $(12x + 3)(2x - 1)$ **32.** $(x - 3)(x - 4)$ **33.** $(2y - 1)(y + 4)$

34. $(5y + 2)(y - 3)$ **35.** $(2x - 3)(3x + 2)$ **36.** $(3x - 2)(x + 2)$

37. $(4x - 5)(x + 5)$ **38.** $(2 - 3x)(7 + x)$ **39.** $(-3 - 4x)(-5 - 6x)$

40. $(1 + 2x)(3 + 4x)$ **41.** $(x - 15)(x - 2)$ **42.** $(y + 10)(y - 3)$

43. $(y + 7)(y - 6)$ **44.** $(x - 8)(x - 7)$ **45.** $(6y + 1)(y - 2)$

46. $(3x - 5)(x + 4)$ **47.** $(2b + 9)(b - 3)$ **48.** $(7 - x)(9 + x)$

Name_____ Class_____ Date_____

Lesson 10-3 Multiplying Polynomials

Multiplying a Trinomial and a Binomial

When multiplying a trinomial and a binomial, use the vertical method or the horizontal method to distribute each term in a factor.

Example

Find the product $(2x^2 - 5x + 7)(3x - 4)$.

Method 1 Multiply vertically.

$$
\begin{array}{r}
2x^2 - 5x + 7 \\
3x - 4 \\
\hline
-8x^2 + 20x - 28 \\
\end{array}
$$
⟵ **Multiply by −4.**

$$6x^3 - 15x^2 + 21x$$
⟵ **Multiply by 3x.**

$$6x^3 - 23x^2 + 41x - 28$$
⟵ **Add like terms.**

Method 2 Multiply horizontally.

$(3x - 4)(2x^2 - 5x + 7)$

$$= 3x(2x^2) + 3x(-5x) + 3x(7) - 4(2x^2) - 4(-5x) - 4(7)$$

$$= 6x^3 - 15x^2 + 21x - 8x^2 + 20x - 28$$

$$= 6x^3 - 23x^2 + 41x - 28$$ ⟵ **Add like terms.**

The product is $6x^3 - 23x^2 + 41x - 28$.

Practice

Multiply.

1. $(3)(x - 3)$

2. $(4)(y + 7)$

3. $(-3)(a + 2)$

4. $(4)(5 - b)$

5. $(4)(x - y)$

6. $(-9)(3 - a)$

Find each product.

7. $(2a^2 + a - 15)(a - 2)$

8. $(x^2 + 2x - 5)(x - 4)$

9. $(-y^2 + 5y + 7)(y + 1)$

10. $(3a^2 - 13a - 10)(a + 3)$

11. $(2c^2 - 3c + 4)(c + 2)$

12. $(x^2 + 2x - 5)(2x - 1)$

13. $(y + 3)(y^2 - 9y - 2)$

14. $(-2 - x)(4x^2 + x - 5)$

15. $(x - 1)(x^2 - 4x + 3)$

16. $(3x + 2)(x^2 - x - 1)$

17. $(y - 9)(2y^2 + y - 2)$

18. $(y + 4)(y^2 - 3y + 5)$

19. $(x + 7)(x^2 - 3x + 4)$

20. $(3y^2 + 2y + 1)(y + 1)$

21. $(9y^2 - 3y + 2)(y - 4)$

22. $(10y^2 - 5y + 1)(y - 5)$

23. $(x^2 - x - 1)(x - 1)$

24. $(x^2 + x + 6)(x + 5)$

© Prentice-Hall, Inc.

Lesson 10-4 Factoring Trinomials

Part **1**

Using Tiles

Example

Use tiles to factor $x^2 - 6x + 9$.

Choose one x^2-tile, six x-tiles and nine 1-tiles. Use the strategy *Guess and Test* to form a rectangle using all the tiles.

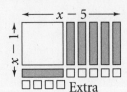

Incorrect

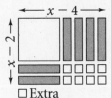

Incorrect

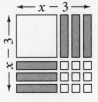

Correct

Write the correct factors as a product.

$x^2 - 6x + 9 = (x - 3)(x - 3)$

Practice

Simplify.

1. $(-3)(-5)(-1)$
2. $(4)(-2)(-2)$
3. $(-3)(3)(-2)$

4. $(-4)(15)(0)$
5. $(-4)(2)(-1)$
6. $(2)(-2)(-2)$

Use tiles or make drawings to represent each expression as a rectangle. Then write the area as the product of two binomials.

7. $x^2 + 8x - 9$
8. $x^2 - x - 6$
9. $x^2 + 6x + 5$

10. $x^2 - 4x + 4$
11. $x^2 - 4x + 3$
12. $x^2 + 6x - 7$

13. $x^2 - x - 2$
14. $x^2 + 6x + 8$
15. $x^2 + 6x + 9$

16. $x^2 - 7x + 12$
17. $x^2 - 5x + 6$
18. $x^2 + x - 20$

19. $x^2 - 8x + 15$
20. $x^2 - 10x + 16$
21. $x^2 + 4x + 3$

22. $x^2 - 2x + 1$
23. $x^2 + 3x + 2$
24. $x^2 + x - 2$

25. $x^2 + 4x + 4$
26. $x^2 + 5x + 4$
27. $x^2 + 2x + 1$

28. $x^2 - 6x + 8$
29. $x^2 + 2x - 15$
30. $x^2 - 3x - 4$

31. $x^2 + 7x - 8$
32. $x^2 + 7x + 10$
33. $x^2 - 11x + 10$

34. $x^2 + 2x - 8$
35. $x^2 - 4x - 12$
36. $x^2 - 6x - 7$

37. $x^2 + 8x + 7$
38. $x^2 - 9x - 10$
39. $x^2 + 9x + 20$

Lesson 10-4 Factoring Trinomials

Testing Possible Factors

To factor trinomials of the form $x^2 + bx + c$, you can use FOIL with the strategy *Guess and Test*.

The sum of the numbers you use here must equal b.

$$x^2 + bx + c = (x \pm \blacksquare)(x \pm \blacksquare)$$

The product of the numbers you use here must equal c.

Example

Factor $x^2 + 10x + 16$.

Choose numbers that are factors of 16. Look for a pair with sum 10.

Factors of 16	Sum of Factors
1 and 16	$1 + 16 = 17$
2 and 8	$2 + 8 = 10$
4 and 4	$4 + 4 = 8$

⟵ List each pair of factors. Just list positive factors because you are looking for a sum of 10. Two negative numbers cannot have a positive sum.

⟵ 2 and 8 have a sum of 10.

The numbers 2 and 8 have a product of 16 and a sum 10. The correct factors are $(x + 2)$ and $(x + 8)$. So, $x^2 + 10x + 16 = (x + 2)(x + 8)$.

Practice

Find the GCF of each pair.

1. $\{9, 12\}$ **2.** $\{4, 16\}$ **3.** $\{6, 15\}$

4. $\{14, 21\}$ **5.** $\{4, 6\}$ **6.** $\{5, 20\}$

Complete.

7. $x^2 - 5x + 6 = (x - 3)(x - \blacksquare)$ **8.** $a^2 + 7a + 12 = (a + 4)(a + \blacksquare)$

9. $b^2 + 4b - 21 = (b + 7)(b - \blacksquare)$ **10.** $y^2 - 2y - 8 = (y - 4)(y + \blacksquare)$

11. $k^2 - 12k + 32 = (k - 8)(k - \blacksquare)$ **12.** $x^2 + 3x - 18 = (x + 6)(x - \blacksquare)$

13. $p^2 - 6p + 5 = (p - 5)(p - \blacksquare)$ **14.** $t^2 + 9t + 18 = (t + 3)(t + \blacksquare)$

15. $y^2 + 7y + 10 = (y + 5)(y + \blacksquare)$ **16.** $t^2 - 8t + 12 = (t - 2)(t - \blacksquare)$

17. $p^2 - 10p + 16 = (p - 2)(p - \blacksquare)$ **18.** $x^2 + x - 30 = (x + 6)(x - \blacksquare)$

19. $y^2 + 9y + 20 = (y + 4)(y + \blacksquare)$ **20.** $t^2 + 5t - 36 = (t + 9)(t - \blacksquare)$

21. $p^2 - 4p - 32 = (p + 4)(p - \blacksquare)$ **22.** $x^2 - 7x - 18 = (x - 9)(x + \blacksquare)$

Lesson 10-4 Factoring Trinomials

Part 3

Factoring Trinomials

To factor quadratic trinomials where $a \neq 1$, list possible factors of a and c.
Use these factors to write binomials. Test for the correct value for b.

Example

Factor $5x^2 + 3x - 8$.

List possible factors of 5: 1 and 5; -1 and -5

List possible factors of -8: 1 and -8; -1 and 8; 2 and -4; -2 and 4

Use the factors to write binomials. Look for 3 as the middle term.

$(1x + 1)(5x + -8)$ $-8x + 5x = -3x$

$(1x + -1)(5x + 8)$ $8x - 5x = 3x$ Correct!

$(1x + 2)(5x + -4)$ $-4x + 10x = 6x$

$(1x + -2)(5x + 4)$ $4x - 10x = -6x$

$5x^2 + 3x - 8 = (x - 1)(5x + 8)$

Practice

List all positive factors of each number.

1. 16   **2.** 21   **3.** 30

4. 14   **5.** 24   **6.** 15

Factor each quadratic trinomial.

7. $2x^2 - 5x - 18$ **8.** $2x^2 + 2x - 12$ **9.** $3x^2 - 7x - 20$

10. $5x^2 + 3x - 14$ **11.** $3x^2 - 4x - 15$ **12.** $6x^2 + 5x - 21$

13. $2x^2 + 4x - 16$ **14.** $7x^2 - 2x - 9$ **15.** $2x^2 + 11x + 14$

16. $4x^2 - 12x + 5$ **17.** $6x^2 + 7x - 20$ **18.** $8x^2 - 13x - 6$

19. $2x^2 + 3x + 1$ **20.** $5x^2 - 7x + 2$ **21.** $6x^2 + 7x - 5$

22. $4x^2 - 9x + 2$ **23.** $12x^2 + 23x - 24$ **24.** $10x^2 + 39x + 36$

25. $12x^2 - 7x + 1$ **26.** $18x^2 - 55x - 28$ **27.** $9x^2 + 25x - 6$

28. $2x^2 + x - 6$ **29.** $3x^2 + 7x + 2$ **30.** $2x^2 - 15x - 77$

31. $15x^2 + 2x - 1$ **32.** $12x^2 - 13x - 35$ **33.** $14x^2 - 13x + 3$

34. $14x^2 + 15x - 9$ **35.** $3x^2 - 7x - 26$ **36.** $4x^2 + 4x + 1$

37. $15x^2 + 26x + 7$ **38.** $5x^2 + 7x - 6$ **39.** $5x^2 + 6x + 1$

40. $3x^2 - 16x + 5$ **41.** $5x^2 - 22x + 8$ **42.** $9x^2 - 30x + 25$

43. $4x^2 - 9x - 9$ **44.** $3x^2 - 13x + 4$ **45.** $6x^2 + 5x + 1$

46. $7x^2 + 47x - 14$ **47.** $6x^2 - 7x - 3$ **48.** $5x^2 + x - 4$

Lesson 10-5 Factoring Special Cases

Part **1**

Factoring a Difference of Two Squares

For all real numbers a and b, $a^2 - b^2 = (a + b)(a - b)$.

Example

Factor $x^2 - 49$.

$\begin{aligned} x^2 - 49 &= x^2 - 7^2 && \longleftarrow \textbf{Rewrite 49 as } 7^2. \\ &= (x + 7)(x - 7) && \longleftarrow \textbf{Factor.} \end{aligned}$

Check.

Use FOIL to multiply.

$(x + 7)(x - 7)$

$x^2 - 7x + 7x - 49$

$x^2 - 49 \checkmark$

So, $x^2 - 49 = (x - 7)(x + 7)$.

Practice

Mini Help

Simplify.

1. $(2x - 3) - (2x - 1)$ **2.** $(3x - 2) + (2 - 3x)$ **3.** $(8b + 3) + (-8b + 5)$

4. $(5 - c) + (6 + c)$ **5.** $(4x - 3) + (3 - 4x)$ **6.** $(5x + 4) + (3 - 5x)$

Factor each expression.

7. $a^2 - 81$ **8.** $x^2 - 16$ **9.** $a^2 - 25$

10. $x^2 - 64$ **11.** $x^2 - 1$ **12.** $b^2 - 36$

13. $4c^2 - 121$ **14.** $9c^2 - 100$ **15.** $x^2 - 9$

16. $y^2 - 169$ **17.** $b^2 - 4$ **18.** $x^2 - 225$

19. $c^2 - 144$ **20.** $4c^2 - 49$ **21.** $16y^2 - 64$

22. $9c^2 - 81$ **23.** $x^2 - y^2$ **24.** $4x^2 - 4$

25. $16x^2 - 9$ **26.** $100x^2 - 25$ **27.** $36y^2 - 16$

28. $49y^2 - 64$ **29.** $121b^2 - 4c^2$ **30.** $4x^2 - 49$

31. $100x^2 - 144$ **32.** $64x^2 - 25$ **33.** $25b^2 - 1$

34. $16x^2 - 81$ **35.** $25x^2 - 100$ **36.** $16x^2 - 25$

37. $144x^2 - 36$ **38.** $b^2 - c^2$ **39.** $49x^2 - 4$

40. $64b^2 - 49$ **41.** $36x^2 - 1$ **42.** $16x^2 - 36$

43. $4c^2 - 144$ **44.** $81x^2 - 100$ **45.** $9c^2 - 16$

46. $25b^2 - 169$ **47.** $144x^2 - 121$ **48.** $36x^2 - 4$

49. $9x^2 - 25$ **50.** $9b^2 - 81$ **51.** $64b^2 - 1$

52. $25y^2 - 9$ **53.** $4b^2 - 1$ **54.** $100x^2 - 400$

© Prentice-Hall, Inc.

Lesson 10-5 Factoring Special Cases

Part **2**

Factoring a Perfect Square Trinomial

For all real numbers a and b:

$$a^2 + 2ab + b^2 = (a + b)(a + b) = (a + b)^2$$
$$a^2 - 2ab + b^2 = (a - b)(a - b) = (a - b)^2$$

Example

Factor $x^2 + 12x + 36$.

$$x^2 + 12x + 36 = x^2 + 12x + 6^2 \quad \longleftarrow \quad \text{Rewrite 36 as } 6^2.$$
$$= x^2 + 2(x)6 + 6^2 \quad \longleftarrow \quad \text{Does the middle term equal } 2ab? \ 12x = 2(x)(6) \ ✔$$
$$= (x + 6)^2 \quad \longleftarrow \quad \text{Factor as a square binomial.}$$

Check.

Use FOIL to multiply.

$(x + 6)(x + 6)$

$x^2 + 6x + 6x + 36$

$x^2 + 12x + 36$ ✔

So, $x^2 + 12x + 36 = (x + 6)^2$.

Practice

Mini ▶ **Help**

Simplify.

1. $\sqrt{100}$ **2.** $-\sqrt{16}$ **3.** $-\sqrt{36}$

4. $\sqrt{1}$ **5.** $\sqrt{9}$ **6.** $-\sqrt{4}$

Factor each expression.

7. $x^2 - 14x + 49$ **8.** $y^2 + 10y + 25$ **9.** $x^2 - 22x + 121$

10. $y^2 + 6y + 9$ **11.** $x^2 - 4x + 4$ **12.** $y^2 + 8y + 16$

13. $y^2 - 20y + 100$ **14.** $x^2 - 24x + 144$ **15.** $x^2 - 16x + 64$

16. $y^2 + 2y + 1$ **17.** $x^2 - 12x + 36$ **18.** $4y^2 - 8y + 4$

19. $25y^2 - 10y + 1$ **20.** $16x^2 - 24x + 9$ **21.** $9x^2 - 30x + 25$

22. $y^2 - 30y + 225$ **23.** $y^2 - 2y + 1$ **24.** $4x^2 - 12x + 9$

25. $x^2 + 4x + 4$ **26.** $9y^2 - 36y + 36$ **27.** $x^2 + 14x + 49$

28. $y^2 + 40y + 400$ **29.** $25x^2 - 60x + 36$ **30.** $4x^2 + 12x + 9$

31. $16y^2 - 8y + 1$ **32.** $9x^2 + 12x + 4$ **33.** $25y^2 - 30y + 9$

34. $y^2 + 20y + 100$ **35.** $4x^2 - 36x + 81$ **36.** $9x^2 - 6x + 1$

37. $x^2 + 22x + 121$ **38.** $4y^2 + 8y + 4$ **39.** $y^2 + 24y + 144$

40. $x^2 - 8x + 16$ **41.** $y^2 - 10y + 25$ **42.** $y^2 - 6y + 9$

Name _____ Class _____ Date _____

Lesson 10-6 Solving Equations by Factoring

When you solve a quadratic equation by factoring, you use the zero product property. For all real numbers a and b, if $ab = 0$, then $a = 0$ or $b = 0$.

Example

Solve $x^2 + x = 20$ by factoring.

$$x^2 + x - 20 = 0 \qquad \longleftarrow \text{ Subtract 20 from both sides.}$$
$$(x - 4)(x + 5) = 0 \qquad \longleftarrow \text{ Factor } x^2 + x - 20.$$
$$x - 4 = 0 \text{ or } x + 5 = 0 \qquad \longleftarrow \text{ Use the zero-product property.}$$
$$x = 4 \text{ or } \qquad x = -5 \qquad \longleftarrow \text{ Solve for } x.$$

The solutions are 4 and -5.

Check. Substitute 4 for x. Substitute -5 for x.

$$(4 - 4)(4 + 5) \stackrel{?}{=} 0 \qquad\qquad (-5 - 4)(-5 + 5) \stackrel{?}{=} 0$$
$$(0)(9) = 0 ✔ \qquad\qquad\qquad (-9)(0) = 0 ✔$$

Practice

Solve each equation for x.

1. $x + 3 = 9$
2. $x - 4 = 24$
3. $x - 5 = 35$
4. $x + 2 = 28$
5. $x - 7 = 42$
6. $x + 9 = 27$

Solve each equation by factoring

7. $a^2 - 5a - 24 = 0$
8. $x^2 - 9x + 18 = 0$
9. $y^2 + 7y + 12 = 0$
10. $b^2 - b - 30 = 0$
11. $c^2 - 12c + 36 = 0$
12. $p^2 + 8p + 12 = 0$
13. $4m^2 - 49 = 0$
14. $k^2 + 16k - 17 = 0$
15. $x^2 - x = 6$
16. $x^2 - 7x = 0$
17. $x^2 + 4x = 12$
18. $x^2 + 7x + 6 = 0$
19. $y^2 - 4y - 12 = 0$
20. $c^2 + 8c + 16 = 0$
21. $x^2 + 2x = 8$
22. $r^2 - r - 90 = 0$
23. $x^2 - 8x + 15 = 0$
24. $n^2 + 5n = 0$
25. $3z^2 - 6z = 0$
26. $4r^2 + 16r = 0$
27. $z^2 - 4z = 21$
28. $2x^2 + 4x = 16$
29. $t^2 - 3t - 18 = 0$
30. $a^2 - 9a - 10 = 0$
31. $x^2 - 4x = 0$
32. $x^2 + 7x = 0$
33. $x^2 - x = 12$
34. $x^2 + 3x = 10$
35. $x^2 + 4x = 32$
36. $x^2 - 7x = 18$
37. $x^2 + 20x = 21$
38. $x^2 - 13x + 40 = 0$
39. $x^2 - 11x + 30 = 0$
40. $x^2 - 5x + 6 = 12$
41. $x^2 - 5x - 24 = 0$
42. $2x^2 - 10x = 0$
43. $a^2 + 25a + 24 = 0$
44. $y^2 - 12y + 27 = 0$
45. $x^2 - 10x + 16 = 0$
46. $t^2 - 9t = -8$
47. $m^2 - 15m + 36 = 0$
48. $a^2 + 17a + 16 = 0$
49. $a^2 = 81$
50. $x^2 - 18x + 81 = 0$
51. $a^2 + 9a + 14 = 0$

© Prentice-Hall, Inc.

Lesson 10-7 Choosing an Appropriate Method for Solving

When you have an equation to solve, first write it in standard form.
Then decide which method to use.

Method	When to use
Graphing	Use if you have a graphing calculator handy.
Square Roots	Use if the equation has only an x^2 term and a constant term.
Factoring	Use if you can factor the equation easily.
Quadratic Formula	Use if you cannot factor the equation or if you are using a scientific calculator.

Example

Solve $2x^2 - 72 = 0$.

Method 1 Square roots

$2x^2 - 72 = 0$

$2x^2 = 72$

$x^2 = 36$

$x = \pm 6$

$x = 6$ or $x = -6$

Method 2 Factoring

$2x^2 - 72 = 0$

$2(x^2 - 36) = 0$

$2(x - 6)(x + 6) = 0$

$x - 6 = 0$ or $x + 6 = 0$

$x = 6$ or $x = -6$

← All methods give →
the same answers.

Method 3 Quadratic Formula

$2x^2 - 72 = 0$

$x = \dfrac{-0 \pm \sqrt{(2)^2 - 4(2)(-72)}}{2(2)}$ ← Use the quadratic formula.

$x = \pm\dfrac{24}{4}$

$x = 6$ or $x = -6$

Practice

Mini ▶ Help

Solve each equation.

1. $2x - 10 = 0$ **2.** $3x + 12 = 0$ **3.** $9x + 18 = 0$

4. $4x + 12 = 0$ **5.** $7x + 14 = 0$ **6.** $5x - 20 = 0$

Solve each equation. Round solutions to the nearest hundredth.

7. $2a^2 + 5a = 12$ **8.** $3y^2 - 8y + 5 = 0$ **9.** $9x^2 + 24x = -16$

10. $5x^2 + 3x - 2 = 0$ **11.** $4a^2 - 64 = 0$ **12.** $4y^2 + 20y = -25$

13. $x^2 + 7x + 6 = 0$ **14.** $4x^2 - 3x - 1 = 0$ **15.** $6x^2 + 4x - 2 = 0$

16. $8x^2 - 32 = 0$ **17.** $9x^2 + 5x - 4 = 0$ **18.** $2y^2 + 4y = -2$

Lesson 11-1 Inverse Variation

Part 1

Solving Inverse Variations

When the product of two quantities remains constant, they form an
inverse variation. As one quantity increases, the other decreases.

Example

The time it takes to walk a certain distance varies inversely with the rate of
speed. Sophia walks at a pace of 3 mi/h. It takes her 2 h to walk home. Her
sister walks at a rate of 4 mi/h. How long does it take her sister to walk home?

Relate Sophia walks 3 mi/h for 2h.

Sophia's sister walks 4 mi/h for x h.

Time varies inversely with rate.

Define $\text{rate}_1 = 3$ mi/h

$\text{rate}_2 = 4$ mi/h

$\text{time}_1 = 2$ h

$\text{time}_2 = x$ h

Write $\text{rate}_1 \cdot \text{time}_1 = \text{rate}_2 \cdot \text{time}_2$ ← **Use a product equation.**

$3 \cdot 2 = 4 \cdot x$ ← **Substitute.**

$6 = 4 \cdot x$

$x = \frac{6}{4}$

$x = 1.5$

It takes Sophia's sister 1.5 h to walk home.

Practice

 Solve.

1. $3x = 24$ **2.** $5y = 25$ **3.** $9g = 81$ **4.** $7m = 49$

Each pair of points is from an inverse variation. Find the missing value.

5. $(4, 10)$ and $(8, x)$ **6.** $(3, 12)$ and $(9, x)$ **7.** $(3, 20)$ and $(4, y)$

8. $(9, 2)$ and $(3, t)$ **9.** $(9, 9)$ and $(27, y)$ **10.** $(10, 10)$ and $(25, c)$

Solve each inverse variation.

11. A trip takes 2 h at 60 m/h. Find the time when the rate is 75 mi/h.

12. How much mass must be placed 2.5 m away from the fulcrum of a lever
in order to balance a 35-kg object that is 3 m from the fulcrum?

13. The length and width of a rectangle of constant area vary inversely. If the
length is 5 when the width is 4, then what is the length when the width is 2?

Lesson 11-1 Inverse Variation
.... Part
Comparing Direct and Inverse Variation **2**

Direct Variation

y varies directly with x.
y is directly proportional to x.
The ratio $\frac{y}{x}$ is constant.

Inverse Variation

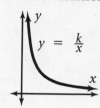

y varies inversely with x.
y is inversely proportional to x.
The product xy is constant.

Example

Are the data in each table a *direct variation* or an *inverse variation*? Write an equation to model the data.

a.

x	6	3	2
y	5	10	15

The values of y seem to vary inversely with the values of x.

Check the product xy.

$xy = 6(5) = 30$

$\quad\quad 3(10) = 30$

$\quad\quad 2(15) = 30$

The product xy is the same for each pair of data. So this is an inverse variation and $k = 30$. The equation is $xy = 30$.

b.

x	3	4	5
y	6	8	10

The values of y seem to vary directly with the values of x.

Check the ratio $\frac{y}{x}$.

$\frac{y}{x} = \frac{6}{3} = 2$

$\frac{8}{4} = 2$

$\frac{10}{5} = 2$

The ratio $\frac{y}{x}$ is the same for each pair of data. So this is a direct variation and $k = 2$. The equation is $y = 2x$.

Practice

Solve each proportion.

1. $\frac{3}{4} = \frac{x}{8}$ **2.** $\frac{6}{4} = \frac{x}{2}$ **3.** $\frac{2}{5} = \frac{x}{15}$ **4.** $\frac{4}{12} = \frac{x}{3}$

Decide whether each table shows a *direct* or an *inverse* variation. Write an equation to model the data.

5.

x	4	3	2
y	6	8	12

6.

x	5	7	9
y	10	14	18

7.

x	4	5	6
y	12	15	18

8.

x	3	4	6
y	12	9	6

9.

x	2	3	5
y	12	18	30

10.

x	2	3	4
y	8	12	16

Lesson 11-2 Rational Functions

Exploring Rational Functions

A **rational function** is a function that can be written in the form
$f(x) = \dfrac{\text{polynomial}}{\text{polynomial}}$.

Example

Evaluate the rational function for $x = 5$.

$$f(x) = \frac{3x - 1}{x^2}$$

$$f(5) = \frac{3(5) - 1}{(5)^2}$$

$$= \frac{15 - 1}{25}$$

$$= \frac{14}{25}$$

Practice

Evaluate for $a = 3, b = 1$.

1. $a + b =$ **2.** $a^2 - b =$ **3.** $(a - b)a =$

4. $b - a =$ **5.** $a + b^2 =$ **6.** $(a + b)^2 =$

Evaluate each function for the given value of x.

7. $y = \dfrac{4}{x - 3}; -3$ **8.** $g(x) = \dfrac{2x}{x + 4}; -3$ **9.** $f(x) = \dfrac{4x}{x^2 + 1}; -3$

10. $y = \dfrac{5 + x}{x}; -3$ **11.** $g(x) = \dfrac{2}{x^2}; -3$ **12.** $f(x) = \dfrac{x^2 - x - 2}{x + 1}; -3$

13. $y = \dfrac{1}{x}; -3$ **14.** $y = \dfrac{x - 1}{x^2 - 5}; -3$ **15.** $g(x) = \dfrac{-1}{x - 5}; 2$

16. $f(x) = \dfrac{-x - 2}{x^2 - 4}; 1$ **17.** $y = \dfrac{x^2 - 3}{2x - 8}; 2$ **18.** $y = \dfrac{x - 1}{1 - x^2}; -2$

19. $f(x) = \dfrac{-x}{x + 3}; 4$ **20.** $g(x) = \dfrac{7}{x^2 + 5}; -3$ **21.** $y = \dfrac{-2x + 10}{x^2 + 3}; -1$

22. $y = \dfrac{-x - 4}{2 - x}; 3$ **23.** $f(x) = \dfrac{3x^2}{2x - 1}; -1$ **24.** $f(x) = \dfrac{2x^2 - 3}{x + 3}; 0$

25. $y = \dfrac{x^2 - 1}{2 - x}; 4$ **26.** $f(x) = \dfrac{x^2 - 7x + 10}{5 - x}; -2$ **27.** $y = \dfrac{x^2 - 5x + 6}{3 - x}; 5$

28. $y = \dfrac{x^2 - 1}{1 - x}; -4$ **29.** $g(x) = \dfrac{x^2 - 36}{-x - 6}; 2$ **30.** $f(x) = \dfrac{2x^2 + x}{x^2 + 3}; 1$

31. $f(x) = \dfrac{3x + 4}{x - 5}; -2$ **32.** $f(x) = \dfrac{9 + x^2}{x^2 - 9}; 4$ **33.** $g(x) = \dfrac{2x^2 + x}{2x + 1}; 5$

34. $g(x) = \dfrac{x^2 - 16}{3x + 12}; 1$ **35.** $y = \dfrac{8x^3 - 2x^2}{4x^2 - x}; 2$ **36.** $y = \dfrac{3x^2 - 7x - 6}{2x^2 - 6x}; 1$

37. $f(x) = \dfrac{10}{x - 6}; 1$ **38.** $f(x) = \dfrac{x + 2}{x^2 - 4}; 1$ **39.** $f(x) = \dfrac{2x^2 + 3}{x - 1}; -2$

Lesson 11-2 Rational Functions

Part 2

Graphing Rational Functions

When you evaluate a rational function, some values of x may lead to division by zero. For the function $y = \frac{1}{x - 1}$, the denominator is zero for $x = 1$.

So, the function is undefined when $x = 1$, and the vertical line $x = 1$ is an asymptote of the graph $y = \frac{1}{x - 1}$.

Example

Graph $y = \frac{1}{x - 1}$.

Step 1 Find the vertical asymptote, which occurs when the denominator equals zero.

$$x - 5 = 0$$

$$x = 5 \quad \longleftarrow \quad \textbf{vertical asymptote}$$

Step 2 Make a table using values of x near 5.

x	y
8	$\frac{1}{3}$
7	$\frac{1}{2}$
6	1
4	-1
3	$-\frac{1}{2}$
2	$-\frac{1}{3}$

Step 3 Draw the graph. **Use a dashed line for the asymptote $x = 5$.**

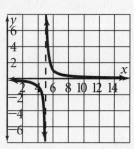

Practice

Solve for x.

1. $x + 12 = 0$
2. $14x = 0$
3. $x^2 + 5x = 0$
4. $x^2 - 9 = 0$
5. $7 - x = 0$
6. $x - 1.8 = 0$

Graph each function. Include a dashed line for each asymptote.

7. $f(x) = \frac{x + 1}{x - 1}$
8. $y = \frac{5}{x - 2} + 1$
9. $y = \frac{3}{x + 2}$
10. $f(x) = \frac{1}{x} - 1$
11. $f(x) = \frac{2}{x} + 3$
12. $y = \frac{8}{x - 3}$
13. $y = \frac{10}{x + 4}$
14. $f(x) = \frac{-2}{x + 1}$
15. $f(x) = \frac{-2}{x}$
16. $f(x) = \frac{6}{x}$
17. $g(x) = \frac{2}{x + 2}$
18. $g(x) = \frac{1}{x} - 1$
19. $y = \frac{1}{x - 4}$
20. $g(x) = \frac{8}{x}$
21. $f(x) = \frac{1}{x - 1}$

Lesson 11-3 Rational Expressions

Part
1

Simplifying Rational Expressions

A **rational expression** is an expression that can be written in the form $\frac{\text{polynomial}}{\text{polynomial}}$, where a variable is in the denominator. The domain is all real numbers excluding the values for which the denominator is zero. The values that are excluded are restricted from the domain. For the expression $\frac{1}{x + 4}$, −4 is restricted from the domain.

Example

Simplify $\frac{3x - 15}{x - 5}$ and state any values restricted from the domain.

$$x - 5 = 0, \quad x = 5 \qquad \longleftarrow \text{ Find the values restricted from the domain.}$$

$$\frac{3x - 15}{x - 5} = \frac{3(x - 5)}{x - 5} \qquad \longleftarrow \text{ Factor the numerator. The denominator cannot be factored.}$$

$$= \frac{3}{1} \cdot \frac{x - 5}{x - 5} \qquad \longleftarrow \text{ Rewrite to show a fraction equal to 1.}$$

$$= 3 \qquad \longleftarrow \text{ Simplify.}$$

The solution is 3. The domain does not include $x = 5$.

Practice

Simplify each expression.

1. $\frac{10x^2}{5}$

2. $\frac{6x}{3}$

3. $\frac{4x}{2x}$

4. $\frac{6x}{11 - 5}$

5. $\frac{5 + x + 2}{3 + 4}$

6. $\frac{x + 1 - x}{4 - 2 - 1}$

Simplify each expression and state any values restricted from the domain.

7. $\frac{b - 2}{b^2 - 4}$

8. $\frac{2y + 6}{y + 3}$

9. $\frac{4b}{2b^2}$

10. $\frac{5x - 20}{3x - 12}$

11. $\frac{x + 1}{x^2 - 1}$

12. $\frac{6a^2}{12a}$

13. $\frac{2x + 4}{2x - 4}$

14. $\frac{x - 1}{2x - 2}$

15. $\frac{x + 4}{x^2 - 16}$

16. $\frac{x}{x^2 - x}$

17. $\frac{2x - 6}{3x - x^2}$

18. $\frac{x^2 - x - 20}{x^2 - 2x - 15}$

19. $\frac{7x + 21}{x + 3}$

20. $\frac{2x - 6}{2x^2 - 5x - 3}$

21. $\frac{12x^2}{6x}$

22. $\frac{3x^2 - 6x}{12 - 6x}$

23. $\frac{2x - 6}{x^2 - 9}$

24. $\frac{2x - 4}{x - 2}$

25. $\frac{5x - 15}{x - 3}$

26. $\frac{x - 1}{1 - x}$

27. $\frac{2x^2 - 5x - 3}{x - 3}$

28. $\frac{3x + 12}{4 + x}$

29. $\frac{x^2 - 5x + 6}{3 - x}$

30. $\frac{x^2 - 1}{1 - x}$

Lesson 11-3 Rational Expressions

Multiplying and Dividing Rational Expressions

To multiply the rational expressions $\frac{a}{b}$ and $\frac{c}{d}$, where $b \neq 0$ and $d \neq 0$, you multiply the numerators and multiply the denominators. Then write the product in simplest form.

$$\frac{a}{b} \cdot \frac{c}{d} = \frac{ac}{bd}$$

To divide the rational expression $\frac{a}{b}$ by $\frac{c}{d}$, where $b \neq 0$, $c \neq 0$, and $d \neq 0$, you multiply by the reciprocal of $\frac{c}{d}$.

$$\frac{a}{b} \div \frac{c}{d} = \frac{a}{b} \cdot \frac{d}{c} = \frac{ad}{bc}$$

Example

Multiply $\dfrac{10x}{9x^2 - 4} \cdot \dfrac{3x - 2}{-5}$.

$$\frac{10x}{9x^2 - 4} \cdot \frac{3x - 2}{-5} = \frac{10x}{(3x - 2)(3x + 2)} \cdot \frac{3x - 2}{-5}$$
← **Factor the denominator. The numerator cannot be factored.**

$$= \frac{\overset{2}{10x}}{\underset{1}{(3x - 2)}(3x + 2)} \cdot \frac{\overset{1}{3x - 2}}{\underset{1}{-5}}$$
← **Divide out the common factors 5 and $(3x - 2)$.**

$$= \frac{-2x}{3x + 2}$$
← **Simplify.**

Practice

Find the reciprocal of each number.

1. $\frac{1}{2}$ 2. $-\frac{1}{5}$ 3. $-\frac{4}{5}$

4. $\frac{3}{5}$ 5. -8 6. $\frac{1}{4}$

Find each product or quotient.

7. $\dfrac{x + 1}{8} \cdot \dfrac{16}{x^2 - 1}$ 8. $\dfrac{7x}{8} \cdot 4$ 9. $\dfrac{2x^2}{3} \div \dfrac{4x^3}{9x}$

10. $\dfrac{6}{5} \cdot \dfrac{25}{18}$ 11. $\dfrac{x^2 - 4}{7} \cdot \dfrac{14}{x - 2}$ 12. $\dfrac{9x}{13} \div \dfrac{1}{26}$

13. $\dfrac{8}{11} \div \dfrac{4}{22}$ 14. $\dfrac{12x^2}{5} \div \dfrac{4x}{5}$ 15. $\dfrac{x}{6} \cdot \dfrac{x}{2}$

16. $\dfrac{10x^2}{8} \div \dfrac{15x^2}{12}$ 17. $\dfrac{4}{x} \cdot \dfrac{x}{2}$ 18. $\dfrac{x^2}{x - 1} \div \dfrac{x}{x - 1}$

19. $\dfrac{3}{x} \div \dfrac{3}{x}$ 20. $\dfrac{3}{x} \div \dfrac{6}{x}$ 21. $\dfrac{5 - 4x}{4} \cdot \dfrac{48}{10 - 8x}$

22. $\dfrac{4x^2}{7} \div \dfrac{2x}{14}$ 23. $\dfrac{12x^2}{6x} \cdot \dfrac{12x}{8x^2}$ 24. $\dfrac{3x^2 - 3}{x + 4} \div \dfrac{x - 1}{2x + 8}$

Name _____ Class _____ Date _____

Lesson 11-4 Operations with Rational Expressions

To add or subtract rational expressions with different denominators, you must write the expressions with a common denominator. Your work will be simpler if you find the least common denominator (LCD), which is the least common multiple of the denominators.

Example

Simplify $\frac{1}{4x} + \frac{5}{8}$.

Step 1 Find the LCD of $\frac{1}{4x}$ and $\frac{5}{8}$.

$4x = 2 \cdot 2 \cdot x$ $\longleftarrow$ **Factor each denominator.**

$8 = 2 \cdot 2 \cdot 2$

The LCD is $2 \cdot 2 \cdot 2 \cdot x$ or $8x$.

Step 2 Write equivalent expressions with denominator $8x$.

$\frac{1}{4x} = \frac{1 \cdot 2}{4x \cdot 2} = \frac{2}{8x}$ $\longleftarrow$ **Multiply numerator and denominator by 2.**

$\frac{5}{8} = \frac{5 \cdot x}{8 \cdot x} = \frac{5x}{8x}$ $\longleftarrow$ **Multiply numerator and denominator by x.**

Step 3 Rewrite the original expression and add.

$\frac{1}{4x} + \frac{5}{8} = \frac{2}{8x} + \frac{5x}{8x}$ $\longleftarrow$ **Replace each expression with its equivalent.**

$= \frac{2 + 5x}{8x}$ $\longleftarrow$ **Add the numerators.**

Practice

Fill in the blank to make each equation true.

1. $4x = \underline{} \cdot x$

2. $3x^2 = 3 \cdot x \cdot \underline{}$

3. $4x = 2 \cdot \underline{} \cdot \underline{}$

4. $8x = 2 \cdot \underline{}$

5. $4x^2 = 2 \cdot x \cdot \underline{}$

6. $-9x = -3 \cdot \underline{} \cdot \underline{}$

Simplify.

7. $\frac{3}{2x} - \frac{5}{4x}$

8. $\frac{4}{5} + \frac{3}{x}$

9. $\frac{7}{2x} + \frac{3}{4}$

10. $\frac{9}{x^2} - \frac{5}{x}$

11. $\frac{5}{2} + \frac{1}{8}$

12. $\frac{2x}{5} + \frac{3}{5}$

13. $\frac{6x}{7} - \frac{1}{7}$

14. $\frac{1}{7} - \frac{x}{2}$

15. $\frac{6}{4b} + \frac{5}{4b}$

16. $\frac{6}{7k} - \frac{5}{7k}$

17. $\frac{7}{5c} - \frac{3}{5c}$

18. $\frac{3}{x} + \frac{4}{x^2}$

19. $\frac{7}{x^2} - \frac{x}{5}$

20. $\frac{1}{x} - \frac{1}{x^2}$

21. $\frac{4}{9} - \frac{5}{3}$

22. $\frac{2}{3} + \frac{4}{12}$

23. $\frac{x}{2} + \frac{3}{4x}$

24. $\frac{1}{2} + \frac{2}{5}$

25. $\frac{3}{x} + \frac{5}{x}$

26. $\frac{7}{2x} - \frac{5}{x}$

27. $\frac{3}{5} + \frac{5}{2}$

Lesson 11-5 Solving Rational Equations

A **rational equation** contains rational expressions. A method for solving rational equations is similar to the method for solving equations with rational numbers.

Example

Solve $\frac{4}{x^2} - 1 = \frac{3}{x}$.

$$x^2\left(\frac{4}{x^2} - 1\right) = x^2\left(\frac{3}{x}\right) \quad \longleftarrow \text{ Multiply each side by the LCD, } x^2.$$

$$x^2\left(\frac{4}{x^2}\right) - x^2(1) = x^2\left(\frac{3}{x}\right) \quad \longleftarrow \text{ Use the Distributive Property.}$$

$$4 - x^2 = 3x \quad \longleftarrow \text{ Simplify.}$$

$$x^2 + 3x - 4 = 0 \quad \longleftarrow \text{ Collect terms on one side.}$$

$$(x + 4)(x - 1) = 0 \quad \longleftarrow \text{ Factor the quadratic expression.}$$

$$(x + 4) = 0 \text{ or } (x - 1) = 0 \quad \longleftarrow \text{ Use the zero-product property.}$$

$$x = -4 \qquad x = 1$$

Check

$$\frac{4}{(-4)^2} - 1 \overset{?}{=} \frac{3}{-4} \qquad\qquad \frac{4}{(1)^2} - 1 \overset{?}{=} \frac{3}{1}$$

$$\frac{4}{16} - 1 = -\frac{3}{4} \qquad\qquad 4 - 1 = 3$$

$$-\frac{3}{4} = -\frac{3}{4} \checkmark \qquad\qquad 3 = 3 \checkmark$$

Since -4 and 1 check in the original equation, they are the solutions.

Practice

Find the sum or difference.

1. $3 - \frac{1}{2}$ 　　　　 **2.** $7 + 1\frac{1}{8}$ 　　　　 **3.** $9 + 2\frac{1}{2}$

4. $1 - \frac{5}{6}$ 　　　　 **5.** $2 - \frac{1}{4}$ 　　　　 **6.** $4 + \frac{10}{11}$

Simplify.

7. $\frac{x}{5} = \frac{3}{5} + \frac{2}{x}$ 　　　 **8.** $3x = \frac{2}{x} + \frac{5}{2}$ 　　　 **9.** $\frac{6}{x^2} - 2 = \frac{4}{x}$

10. $\frac{2}{x^2} - 3 = \frac{1}{x}$ 　　 **11.** $\frac{1}{x} + \frac{1}{2} = \frac{4}{x^2}$ 　　 **12.** $\frac{3}{4} + \frac{1}{3x} = 1$

13. $\frac{x}{2} + \frac{5}{x} = \frac{3}{2}$ 　　 **14.** $x + \frac{1}{2} = \frac{x}{3}$ 　　 **15.** $1 - \frac{5}{x + 7} = \frac{2x}{x + 7}$

16. $\frac{8}{2x - 1} = 4 - \frac{4x}{2x - 1}$ 　　 **17.** $1 - \frac{3}{x} = 4$ 　　 **18.** $7 + \frac{6}{x} = 5$

19. $\frac{9}{x} + 1 = \frac{12}{x}$ 　　 **20.** $\frac{3}{x} - 2 = -\frac{21}{x}$ 　　 **21.** $1 + \frac{2}{x + 1} = \frac{3}{x + 1}$

Lesson 11-6 Counting Outcomes and Permutations

Part 1

Using the Multiplication Counting Principle

If there are m ways to make a first selection and n ways to make a second selection, there and $m \times n$ ways to make the two selections.

Example

In a dressing room, there are seven shirts and five pairs of pants. Find the number of possible outfits.

$$7 \cdot 5 = 35 \quad \longleftarrow \quad \textbf{number of possible outfits}$$

number of shirts ↑ ↑ **number of pants**

Practice

Find each product.

1. $3 \cdot 5$ **2.** $8 \cdot 4$ **3.** $11 \cdot 10$

4. $6 \cdot 7$ **5.** $13 \cdot 9$ **6.** $2 \cdot 12$

Simplify.

7. A code must consist of one letter from the alphabet and one whole number that is less than 10. How many possible codes are there?

8. Suppose there are two routes from downtown to your school, and three routes from your school to the airport. How many routes are there from downtown to the airport through your school?

9. In a cabinet there are four cans of fruit and seven cans of vegetables. In how many ways can you select one of each?

10. In an animal shelter, there are 12 cats and 14 dogs. How many ways can you adopt one of each?

11. In a waiting room, there are five different magazines and four different newspapers. How many ways can you choose one of each?

12. At a pizza shop, there are seven different thin-crust pizzas and eight different thick-crust pizzas. How many ways can you order one of each?

13. In a classroom, there are 20 students and 15 teachers. How many possible teacher-student pairs are there?

14. At a restaurant there are 11 colors of napkins and 11 types of napkin rings. How many ways can you pair a napkin with a napkin ring?

Lesson 11-6 Counting Outcomes and Permutations

Finding Permutations

A *permutation* is an arrangement of some or all of a set of objects in a specific order. You can use the multiplication counting principle to evaluate permutations.

$$\underset{\uparrow\text{ number of factors}}{{}_nP_r} = \overset{\downarrow\text{ first factor}}{n}(n-1)(n-2)...$$

Example

A four letter identification code is needed to check out books from the school library. Find the possible number of four-letter codes.

$$_{26}P_4 = 26 \cdot 25 \cdot 24 \cdot 23$$

$$= 358,800 \qquad \longleftarrow \textbf{Use a calculator.}$$

There are 358,800 four-letter codes in which letters do not repeat.

Practice

Evaluate.

1. $5 \cdot 4 \cdot 3$ **2.** $7 \cdot 6 \cdot 5$ **3.** $10 \cdot 9 \cdot 8$

4. $12 \cdot 11 \cdot 10$ **5.** $4 \cdot 3 \cdot 2$ **6.** $6 \cdot 5 \cdot 4$

Use a calculator to solve.

7. Toni needs to pick a 9-digit security code for her new apartment. She cannot use the same number twice. How many different codes can she have?

8. How many ways can you arrange the letters in the word TEXAS?

9. How many ways can a teacher call on seven students to give their answer?

10. How many ways can six students stand in line at a water fountain?

11. There are ten finalists in a competition. How many ways can first, second, and third place be awarded?

12. There are five seats in the front row of a classroom of 15 students. How many ways can students be assigned to sit in the front row?

13. $_7P_2$ **14.** $_{10}P_2$ **15.** $_5P_4$

16. $_4P_3$ **17.** $_4P_4$ **18.** $_5P_3$

19. $_{10}P_3$ **20.** $_7P_3$ **21.** $_6P_3$

Lesson 11-7 Combinations

$$\text{number combinations} = \frac{\text{total number of permutations}}{\text{number of times the objects in each group are represented}}$$

In the notation ${}_nC_r$, the number of combinations is shown with n representing objects chosen r at a time. The number of times each group of objects is repeated depends on r.

$${}_nC_r = \frac{{}_nP_r}{{}_rP_r} = \frac{n(n-1)(n-2)\dots}{r(r-1)(r-2)\dots}$$

Example

Fifteen students try out for a ten-member team. How many different teams can be chosen?

The order in which team members are listed does not distinguish one team from another. So find the number of combinations of 15 students chosen 10 at a time.

$${}_{15}C_{10} = \frac{15 \cdot 14 \cdot 13 \cdot 12 \cdot 11 \cdot 10 \cdot 9 \cdot 8 \cdot 7 \cdot 6}{10 \cdot 9 \cdot 8 \cdot 7 \cdot 6 \cdot 5 \cdot 4 \cdot 3 \cdot 2 \cdot 1}$$

$$= 3003 \qquad \longleftarrow \textbf{Use a calculator.}$$

There are 3003 different ten member teams possible.

Practice

Evaluate.

1. ${}_5P_2$ **2.** ${}_7P_3$ **3.** ${}_{12}P_5$

4. ${}_4P_1$ **5.** ${}_{10}P_4$ **6.** ${}_6P_2$

Use a calculator to solve.

7. There are 12 fluttering butterflies. A biology student's net can catch only seven butterflies at a time. How many different seven-butterfly groups can the student catch to observe?

8. Pat has seven differently colored pencils. For the exam, students are allowed to have only four pencils. How many different four-pencil combinations can Pat choose?

9. There are five finalists in a contest. Only three will be given medals. How many different combinations of medal recipients are there?

10. Alex is preparing a salad with four of these ingredients: lettuce, tomatoes, cucumbers, croutons, celery, cauliflower, and carrots. How many different four-ingredient salad combinations are there?

11. Six friends go to an amusement park. They want to sit in the front row of a ride that holds only four people per row. How many ways can the friends choose who will sit in the front row?

Two-Year Algebra Handbook Answers

1-1 ▼ Finding Mean, Median, and Mode

1. 1 2 3 4 5 6 7 9 **2.** −2 0 1 2 3 3 4 5
3. −23 −15 −14 −6 −4 **4.** 8.5 17 20 47.5 65
5. 0.9 1.0 1.3 1.5 1.7 **6.** −13 2.3 31 50.5 75
7. mean = 3.75; median = 3.5; mode = 3
8. Height of Players on the Girls Basketball Team

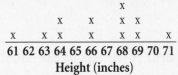

              x
        x   x   x   x
    x   x   x   x   x x
   61 62 63 64 65 66 67 68 69 70 71
            Height (inches)

9. mean ≈ 66; median = 67; mode = 68 **10.** Mode and median decrease to 66; mean remains about 66.

1-1 ▼ Drawing and Interpreting Graphs

1. South **2.** North **3.** West
4. Midwest and South
5a. Answers may vary. Sample:

U.S. Passenger Car Production

(graph: Cars produced (millions) vs. Year 1970–1990; General Motors ♦, Ford ■, Chrysler ▲)

5b. Answers may vary. Sample: Trends are more easily seen on a multiple line graph. **5c.** Ford and Chrysler production declined every period except 1980 to 1985; General Motors production increased every period except 1985 to 1990.
5d. Production declined at all three manufacturers between 1985 and 1990.

1-2 Modeling Relationships with Variables

1. 8, 10, 12 **2.** 12, 15, 18 **3.** 12, 16, 20 **4.** −9, −11, −13
5. 16, 32, 64 **6.** 8, 5, 2 **7.** $d = 10m$ **8.** $4.75h = 64$
9. $c = 30d$ **10.** $p = 4m$ **11.** $m = 150w$
12. $l = 15 - u$

1-3 ▼ Evaluating Expressions

1. 12 **2.** −12 **3.** 30 **4.** 18 **5.** −9 **6.** 12 **7.** 7 **8.** 8
9. 44 **10.** 9 **11.** 1 **12.** 196 **13.** 26 **14.** −1 **15.** 11
16. 24 **17.** 17 **18.** −6 **19.** 112 **20.** 4.6
21. about 2,415,766.7 m^3 **22.** 4186.67 cubic km

1-3 ▼ Evaluating Expressions With Grouping Symbols

1. 14 **2.** 67 **3.** $\frac{4}{7}$ or 0.57 **4.** 16 **5.** 63 **6.** 37.6 **7.** 30
8. 1 **9.** 36 **10.** 20 **11.** −17 **12.** $\frac{9}{8}$ **13.** −6 **14.** 2
15. 4 **16.** 64 **17.** 23 **18.** 10 **19.** 0 **20.** $\frac{25}{6}$ or 4.17
21. about 84 m^3 **22.** $300

1-4 ▼ Adding Integers and Decimals

1. 9 **2.** 4 **3.** −4 **4.** −6 **5.** −2 **6.** 0 **7.** −4 **8.** −3 **9.** −5
10. 3 **11.** 2 **12.** 5 **13.** −9 **14.** 0 **15.** −13°F **16.** 37°F

1-4 ▼ Subtracting Integers and Decimals

1. 8 **2.** 7 **3.** 19 **4.** 3 **5.** 14 **6.** 1.2 **7.** −2 **8.** −7 **9.** 2
10. 4 **11.** −2 **12.** 7 **13.** −7 **14.** 7 **15.** −10 **16.** 10
17. 16 **18.** 13 **19.** 7 **20.** 3 **21.** −12 **22.** 18 **23.** −4
24. 0 **25.** 1 **26.** 6 **27.** 0 **28.** 1 **29.** −9 **30.** 7

1-5 ▼ Multiplying and Dividing

1. 2 **2.** −12 **3.** 8 **4.** −29 **5.** −4 **6.** −13 **7.** −30 **8.** −4
9. 2 **10.** 12 **11.** −12 **12.** −$\frac{1}{4}$ **13.** −5 **14.** 9 **15.** $60
16a. $d = 550 - 4.5w$ **16b.** 58.5 ft **16c.** 491.5 ft

1-5 ▼ Simplifying Expressions with Exponents

1. 27 **2.** 16 **3.** 32 **4.** 125 **5.** −216 **6.** −16 **7.** −27
8. −49 **9.** −27 **10.** −27 **11.** −16 **12.** 16 **13.** −25
14. 32 **15.** 1 **16.** 49 **17.** −125 **18.** −25 **19.** −64
20. −8 **21.** −16 **22.** 4 **23.** 625 **24.** 36 **25.** −64 **26.** −81
27. 81 **28.** 36 **29.** 81 **30.** −16 **31.** 27 **32.** 9 **33.** 64
34. −8 **35.** 49 **36.** −64 **37.** −125 **38.** −27 **39.** −2401

1-6 ▼ Real and Rational Numbers

1. < **2.** < **3.** > **4.** = **5.** = **6.** >
7. $-\frac{3}{4}, -\frac{7}{12}, -\frac{8}{15}$ **8.** $\frac{3}{8}, \frac{5}{12}, 0.43$ **9.** $-\frac{13}{25}, -\frac{6}{13}, -\frac{2}{4}$
10. −0.37, −0.35, −0.037 **11.** $-\frac{3}{11}, -\frac{3}{13}, \frac{2}{12}$
12. $\frac{9}{10}, \frac{10}{11}, \frac{11}{12}$ **13.** −0.55, $-\frac{7}{13}$, −0.5
14. $-\frac{7}{9}, -\frac{5}{7}, -\frac{3}{5}$ **15.** $-\frac{1}{2}, -\frac{1}{3}, -\frac{1}{4}$
16. $-\frac{1}{7}$, −0.125, $-\frac{1}{9}$ **17.** $-\frac{4}{5}, -\frac{2}{3}, -\frac{9}{16}$
18. −4.5, −0.45, −0.0045 **19.** −0.96, −0.45, 67
20. $\frac{9}{11}, \frac{10}{12}, \frac{9}{10}$ **21.** $-\frac{12}{9}, \frac{10}{9}, \frac{11}{9}$
22. $-\frac{3}{2}, -\frac{9}{18}, -\frac{10}{50}$ **23.** $\frac{1}{3}, \frac{10}{13}, \frac{36}{40}$
24. −0.648, −0.537, −0.058

1-6 ▼ Evaluating Expressions

1. 6 **2.** 8 **3.** 12 **4.** −6 **5.** 60 **6.** 20 **7.** $\frac{31}{12}$ or 2.58

8. $\frac{13}{6}$ or $2\frac{1}{6}$ **9.** $-\frac{5}{4}$ **10.** 6 **11.** $8\frac{1}{3}$ **12.** $6\frac{1}{4}$ **13.** 0

14. 5 **15.** $\frac{1}{24}$ **16.** $\frac{8}{7}$ **17.** 2 **18.** $\frac{17}{15}$ **19a.** $1\frac{2}{5}$ in.

19b. $\frac{13}{20}$ in. **19c.** $1\frac{11}{20}$ in. **20a.** 8 km **20b.** $\frac{28}{5}$

20c. $\frac{8}{15}$ km

1-7 ▼ Finding Experimental Probability

1. 50% **2.** 48% **3.** 65.2% **4.** 38% **5.** 85.7% **6.** 90.8%

7. 3.2% **8.** 47.9% **9.** 13.8% **10.** 17% **11.** $\frac{15}{63}$, 23.8%

12. $\frac{4}{63}$, 6.3% **13.** $\frac{11}{63}$, 17.5% **14.** $\frac{9}{63}$, 14.3%

1-7 ▼ Conducting a Simulation

1. 20% **2.** 16.7% **3.** 60% **4.** 60% **5.** 64.3% **6.** 16.7%
7–13. Answers may vary.

1-8 Organizing Data in Matrices

1. −37 **2.** 5 **3.** 12 **4.** 0 **5.** −9 **6.** 14

7. Addition Subtraction

$$\begin{bmatrix} 9 & 10 \\ 18 & 12 \end{bmatrix}, \begin{bmatrix} -1 & 6 \\ 2 & 6 \end{bmatrix}$$

8. Addition Subtraction

$$\begin{bmatrix} 7 & 17 & 18 \\ 5 & 18 & 11 \end{bmatrix}, \begin{bmatrix} 3 & 5 & 0 \\ 3 & 2 & 3 \end{bmatrix}$$

9. Addition Subtraction

$$\begin{bmatrix} 14 & 5 \\ 10 & 11 \\ 6 & 9 \end{bmatrix}, \begin{bmatrix} -4 & -1 \\ 6 & 3 \\ -4 & 9 \end{bmatrix}$$

10. Addition Subtraction

$$\begin{bmatrix} 3 & 5 \\ 7 & 10 \end{bmatrix}, \begin{bmatrix} -1 & -5 \\ -7 & -8 \end{bmatrix}$$

11. Addition Subtraction

$$\begin{bmatrix} 3 & 8 & 5 \\ 3 & 9 & 13 \\ 16 & 3 & 6 \end{bmatrix}, \begin{bmatrix} 1 & -2 & -3 \\ -1 & -3 & -1 \\ -2 & 1 & 2 \end{bmatrix}$$

12. Addition Subtraction

$$\begin{bmatrix} 10 & 9 & 13 \\ 6 & 5 & 10 \end{bmatrix}, \begin{bmatrix} -6 & -7 & -3 \\ 0 & -5 & 8 \end{bmatrix}$$

13.
$$\begin{bmatrix} 560 & 139 & 118 \\ 720 & 180 & 150 \\ 547 & 160 & 151 \end{bmatrix}$$

1-9 Variables and Formulas in Spreadsheets

1. 120 **2.** 2 **3.** 27 **4.** 15 **5.** 2 **6.** 14
7. $4*A2 - 5, A2^2 + 1, A2 - 6$ **8.** 31, 82, 3 **9.** 7, 10, −3

2-1 ▼ Drawing and Interpreting Scatter Plots

1–6.

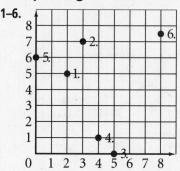

7.

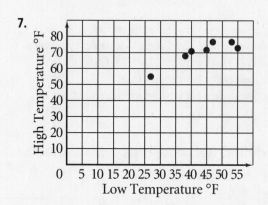

8. about $65 **9.** about 80 oz **10.** no

2-1 ▼ Analyzing Trends Data

For exercises **1–4**, answers may vary. Samples:

1. 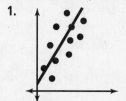 **2.**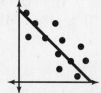

3.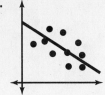

4.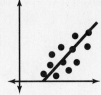

5. Answers may vary. Sample:

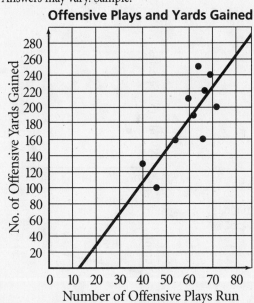

6. positive correlation   **7.** approximately 260 yd

2-2 ▼ Interpreting Graphs

1. increasing   **2.** decreasing   **3.** staying the same
4. increasing   **5.** decreasing

6.

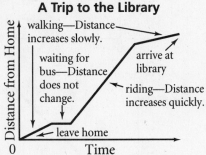

7.

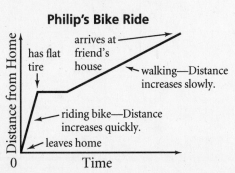

2-2 ▼ Sketching Graphs

1.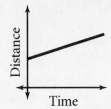

2.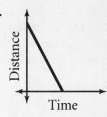

3.

4.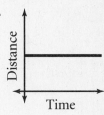

5.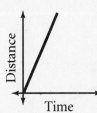

6. Answers may vary. Sample:

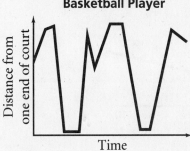

7. Answers may vary. Sample:

Two-Year Algebra Handbook Answers (continued)

2-2 ▼ Classifying Data

1. counted **2.** measured **3.** measured **4.** counted
5. measured **6.** counted
7–10. Graphs may vary. Samples included.
7. discrete

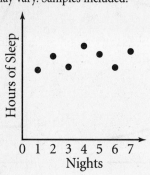

8. continuous

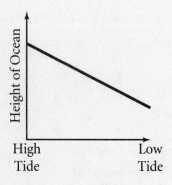

9. continuous

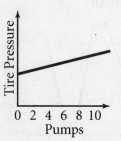

10. continuous

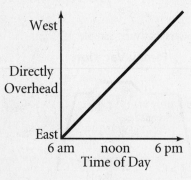

11. discrete

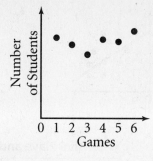

12. discrete

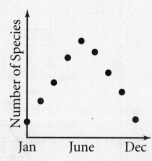

13. continuous

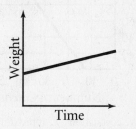

14. discrete

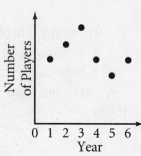

15. continuous

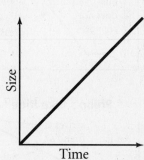

16. discrete

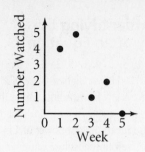

2-3

For exercises **1–4**, answers may vary. Sample:

1.
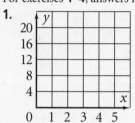

2.
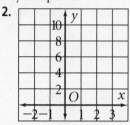

3.
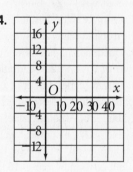

4.

5.
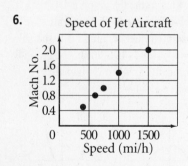

Pizza Menu

6. Speed of Jet Aircraft

7. Bacterial Reproduction
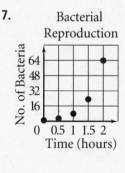

2-4 ▼ Identifying Relations and Functions

1. $\{(0, 4), (-5, 0), (13, 6), (7, -11)\}$ **2.** $\{(20, 24), (33, -15),$ $(57, 14), (19, 8)\}$ **3.** $\{(0.50, 17), (0.75, 11), (0.25, 13), (1.15, 9)\}$ **4.** $\{(76, 115), (0, -70), (113, -12), (-84, 0)\}$
5. function **6.** not a function **7.** function
8. not a function **9.** not a function **10.** function; ocean depth is independent variable, pressure is dependent variable
11. not a function **12.** function; number of correct answers is independent variable, test grade is dependent variable

2-4 ▼ Evaluating Functions

1. 7 **2.** -13 **3.** -5 **4.** -9 **5.** 47 **6.** -2.5 **7.** 5 **8.** $-\frac{1}{2}$
9. 29 **10.** 14 **11.** -25 **12.** 3 **13.** 3 **14.** 11 **15.** -9
16. 5 **17.** 20 **18.** -53 **19.** -1 **20.** -14 **21.** $\{-15, 0, 25\}$
22. $\{-13, -4, 11\}$ **23.** $\{8, 5, 0\}$ **24.** $\{1, \frac{5}{2}, 5\}$
25. $\{1, 10, -15\}$ **26.** $\{-1, -4, \frac{13}{3}\}$ **27.** $\{19, -8, -133\}$
28. $\{10.4, 1.4, 26.4\}$ **29.** $\{-3, 3, 11\}$ **30.** $\{-13, -4, 8\}$
31. $\{-\frac{7}{4}, -1, -2\}$ **32.** $\{0, 1, \frac{7}{3}\}$ **33.** $\{10, 4, -4\}$
34. $\{-12, 0, 16\}$ **35.** $\{\frac{7}{3}, \frac{1}{3}, -\frac{7}{3}\}$
36. $\{-11.1, -2.1, -18.1\}$

2-4 ▼ Analyzing Graphs

1. horizontal **2.** vertical **3.** neither **4.** vertical **5.** neither
6. horizontal **7.** yes **8.** yes **9.** yes **10.** no **11.** yes **12.** no
13. yes **14.** yes

2-5 ▼ Understanding Function Notation

1. 13 **2.** -20 **3.** -17 **4.** -32 **5.** 10 **6.** $\frac{1}{3}$
7. $f(x) = -2x + 5$ **8.** $f(x) = 7 - x^2$
9. $f(x) = 0.3x + 1.6$ **10.** $f(x) = 4x^3 - 57$ **11.** 7
12. 16 **13.** -8 **14.** -22 **15.** -2 **16.** 17 **17.** 0 **18.** 28
19. $\{13, 1, -23\}$ **20.** $\{-4, 5, 23\}$ **21.** $\{9, 3, 99\}$
22. $\{-7, -\frac{17}{2}, -\frac{23}{2}\}$ **23.** $\{2, 5, 11\}$ **24.** $\{-19, -1, 35\}$
25. $\{-28, -7, -343\}$ **26.** $\{2.1, 9.3, 23.7\}$ **27.** $\{-9, 1, 21\}$
28. $\{0, -1, 1\}$ **29.** $\{-6, -2, 6\}$ **30.** $\{7, 4, -2\}$ **31.** $\{4, 1, 13\}$
32. $\{3.4, 4.7, 7.3\}$ **33.** $\{-3, 0, -12\}$ **34.** $\{-\frac{9}{2}, -4, -3\}$

2-5 ▼ Using A Table of Values

1. Answers may vary. Sample: $f(x) = x + 2, f(x) = 3x$
2. Answers may vary. Sample: $f(x) = -x, f(x) = x - 4$
3. Answers may vary. Sample: $f(x) = x - 1, f(x) = \frac{4}{5}x$
4. Answers may vary. Sample: $f(x) = 2x + 3, f(x) = x^2$
5. Answers may vary. Sample: $f(x) = 1 - x,$
$f(x) = 0.5x - 8$

6. Answers may vary. Sample: $f(x) = 2x - 1, f(x) = 6 + x$
7. C **8.** A **9.** B **10.** $f(x) = x + 3$ **11.** $f(x) = x - 5$
12. $f(x) = \frac{1}{3}x$ **13.** $f(x) = -2x$

2-5 ▼ Using Words to Write a Rule

1. independent variable: gallons of gasoline; dependent variable: cost of filling tank **2.** independent variable: position of moon; dependent variable: height of tide **3.** independent variable: amount of deposit; dependent variable: balance of account **4.** $C(s) = 5 - 0.32s$ **5.** $I(c) = 0.05c$
6. $P(p) = p - 500$ **7a.** $K(m) = 1.6m$ **7b.** 16
8a. $P(b) = 1.5b - 25$ **8b.** 17

2-6

1. 3; 11 **2.** 32; -16 **3.** -10; 14 **4.** -3; 1 **5.** 1; 7 **6.** 10; 14

7a. Answers may vary. Sample:

t	$D(t) = 330 - 55t$	(t, D(t))
1	$330 - 55(1) = 275$	(1, 275)
3	$330 - 55(3) = 165$	(3, 165)
4	$330 - 55(4) = 110$	(4, 110)
5	$330 - 55(5) = 0$	(5, 55)

7b. **Train Distance** **7c.** 8 P.M.

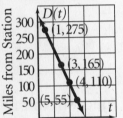

8–10. Answers may vary.
8. Sample:

x	y
1	4
0	2
1	0
2	-2
3	-4

9. Sample:

x	y
-2	5
0	-1
1	-4
2	-1
3	2

10. Sample:

x	y
-4	1
-2	4
0	5
2	4
6	-4

2-7 ▼ Identifying the Family of an Equation

1. 2 **2.** 1 **3.** 2 **4.** 1 **5.** 2 **6.** 3 **7.** Quadratic; highest power of x is 2. **8.** Linear; highest value of x is 1. **9.** Linear; highest power of x is 1. **10.** Absolute value; absolute value symbol around variable expression. **11.** Quadratic; highest power of x is 2. **12.** Linear; highest power of x is 1.
13. Absolute value; absolute value symbol around variable expression. **14.** Quadratic; highest power of x is 2.
15. Absolute value; absolute value symbol around variable expression. **16.** Linear; highest power of x is 1.
17. Quadratic; highest power of x is 2. **18.** Quadratic; highest power of x is 2. **19.** Quadratic; highest power of x is 2.
20. Absolute value; absolute value symbol is around variable. **21.** Quadratic; highest power of x is 2. **22.** Linear; highest power of x is 1. **23.** Absolute value; absolute value symbol is around variable. **24.** Linear; highest power of x is 1.
25. Quadratic; highest power of x is 2. **26.** Linear; highest power of x is 1. **27.** Linear; highest power of x is 1.
28. Absolute value; absolute value symbol is around variable. **29.** Absolute value; absolute value symbol is around variable. **30.** Quadratic; highest power of x is 2. **31.** Quadratic; highest power of x is 2. **32.** Quadratic; highest power of x is 2.
33. Linear; highest power of x is 1. **34.** Linear; highest power of x is 1. **35.** Absolute value; absolute value symbol is around variable. **36.** Absolute value; absolute value symbol is around variable. **37.** Quadratic; highest power of x is 2. **38.** Linear; highest power of x is 1. **39.** Absolute value; absolute value symbol is around variable. **40.** Answers may vary. Sample:

linear: $y = 2x, y = 4 - 5x, y = -\frac{3}{2}x + 1$; quadratic:
$y = -x^2, y = 4x^2 + 3, y = 11 - \frac{1}{4}x^2$; absolute value:
$y = |6x|; y = |2 - 3x|, y = 1 - \left|\frac{2}{3}x\right|$

2-7 ▼ Identifying the Family of a Graph

1. V-shaped **2.** straight line **3.** U-shaped **4.** Linear; the graph is a straight line. **5.** Quadratic; the graph is U-shaped.
6. Absolute value; the graph is V-shaped. **7.** Quadratic; the graph is U-shaped. **8.** Absolute value; the graph is V-shaped.
9. Linear; the graph is a straight line. **10.** Linear; the graph is a straight line. **11.** Quadratic; the graph is U-shaped.
12. Absolute value; the graph is V-shaped.
13. Answers may vary. Sample:

Two-Year Algebra Handbook Answers (continued)

2-8 ▼ Finding Theoretical Probability

1. $\frac{1}{4}$ 2. $\frac{1}{5}$ 3. $\frac{3}{4}$ 4. $\frac{2}{5}$ 5. $\frac{17}{20}$ 6. $\frac{3}{5}$ **7a.** 5 **7b.** 20% **7c.** 80% **8a.** 50% **8b.** 0% **8c.** 70% **8d.** 50%

2-8 ▼ Using a Tree Diagram to Find a Sample Space

1.   2.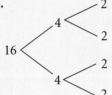

3. 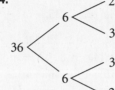 4. 5. 6.

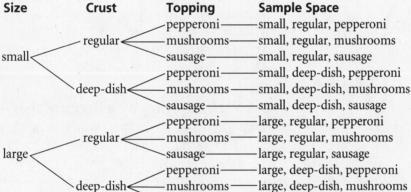

7a.

Size	Crust	Topping	Sample Space
small	regular	pepperoni	small, regular, pepperoni
		mushrooms	small, regular, mushrooms
		sausage	small, regular, sausage
	deep-dish	pepperoni	small, deep-dish, pepperoni
		mushrooms	small, deep-dish, mushrooms
		sausage	small, deep-dish, sausage
large	regular	pepperoni	large, regular, pepperoni
		mushrooms	large, regular, mushrooms
		sausage	large, regular, sausage
	deep-dish	pepperoni	large, deep-dish, pepperoni
		mushrooms	large, deep-dish, mushrooms
		sausage	large, deep-dish, sausage

7b. 17%

8a.

Left	Middle	Right	Sample Space
Erik	Tina	Elisa	Erik, Tina, Elisa
	Elisa	Tina	Erik, Elisa, Tina
Tina	Erik	Elisa	Tina, Erik, Elisa
	Elisa	Erik	Tina, Elisa, Erik
Elisa	Erik	Tina	Elisa, Erik, Tina
	Tina	Erik	Elisa, Tina, Erik

8b. 33% **8c.** 67%

3-1 ▼ Solving Addition and Subtraction Equations

1. 9.6 **2.** −5 **3.** −25 **4.** $\frac{9}{8}$ **5.** 8.8 **6.** $1\frac{2}{3}$ **7.** 16

8. 4.5 **9.** 18 **10.** 37 **11.** $\frac{1}{2}$ **12.** −8 **13.** 6.9 **14.** −7

5. 0 **16.** 17 **17.** −4 **18.** 53

3-1 ▼ Solving Multiplication and Division Equations

1. −5.4 **2.** 1.8 **3.** $\frac{1}{3}$ **4.** 3.0 **5.** $-\frac{3}{4}$ **6.** $-\frac{1}{4}$ **7.** −13

8. −9 **9.** 17.0 **10.** 44 **11.** −19.5 **12.** 324 **13.** −60

14. $-\frac{3}{8}$ **15.** −3 **16.** 0 **17.** 1 **18.** 10 **19.** −8 **20.** −3

21. 7 **22.** 4 **23.** −6 **24.** 55 **25.** $-\frac{9}{2}$ **26.** 148.5 **27.** −45

3-1 ▼ Modeling by Writing Equations

1. $x + 5$ **2.** $a - 15$ **3.** $8k$ **4.** $\frac{n}{2.8}$ **5.** 1046 ft

6. 7 million **7.** $106.25 **8.** 14 problems **9.** $24.00

3-2 ▼ Using Tiles

1. −2 **2.** −36.5 **3.** 11 **4.** 3 **5.** 1 **6.** 2 **7.** 3 **8.** 4 **9.** 1
10. 4 **11.** −1 **12.** −1 **13.** −6 **14.** 1 **15.** 4 **16.** −2
17. $\frac{8}{3}$ **18.** 31 **19.** 3

3-2 ▼ Using Properties

1. 2 **2.** 6 **3.** −6 **4.** −35 **5.** 16 **6.** $\frac{5}{3}$ **7.** 5 **8.** 2 **9.** 2

10. 0 **11.** −17 **12.** 38.4 **13.** −19 **14.** 0.9 **15.** 20 **16.** 10
17. 1 **18.** 56 **19.** $18 **20.** 4 pair **21.** 3 **22.** $4.65

3-3 ▼ Combining Like Terms

1. −13 **2.** −7 **3.** 9 **4.** 10 **5.** 4 **6.** 0 **7.** $5c$ **8.** −5m
9. $2 - 4y$ **10.** $8a$ **11.** $3x - 4$ **12.** $-b - 7$ **13.** $4 - k$
14. $11 - 5z$ **15.** 0 **16.** $-k - m$ **17.** $a - 6$
18. $t - 4r$ **19.** $-5e - f$ **20.** $-6x - 6y + 8z - 6$
21. $2a - b$ **22.** $-0.9c - d$ **23.** $8x$ **24.** $2x + y + 3$
25. $7x + 7x^2$ **26.** $4m + 13$ **27.** $3s + 4$
28. $7y - 4x - 4$ **29.** $5x^2 + 15$ **30.** $10m - 22n + 8$
31. $b^2 + 20b$ **32.** $12g - h + 2$

3-3 ▼ Solving Equations

1. $-13a$ **2.** $-7x$ **3.** $14b$ **4.** $8y$ **5.** $4a$ **6.** 0 **7.** 6 **8.** −3

9. 6 **10.** $\frac{4}{9}$ **11.** −0.96 **12.** 6 **13.** −3.08 **14.** −3

15. 1.25 in. **16.** $11.00 **17.** 5, 6, 7, and 8 yr **18.** 11 cm

3-4 ▼ Simplifying Variable Expressions

1. 24 **2.** −30 **3.** −55 **4.** 54 **5.** −220 **6.** 169 **7.** −3
8. −8 **9.** $8x - 24$ **10.** $3a - 21$ **11.** $-8n - 40$
12. $-6b - 2$ **13.** $-3t + 7$ **14.** $-p - 8$
15. $24r - 72$ **16.** $3x + 21$ **17.** $-12 + 10k$
18. $3w - 9$ **19.** $\frac{1}{2}g + 4$ **20.** $12d - 20$ **21.** $16n - 8$
22. $9x - 3$ **23.** $48t - 56$ **24.** $-4p + 5$
25. $-12a - 4.5$ **26.** $-35 + 15y$ **27.** $-3a - 6$
28. $-12y$ **29.** $6x - 22$ **30.** $14y + 42$
31. $-\frac{1}{2}x + 6$ **32.** $-25j + 100$ **33.** $12x - 6$
34. $4a + 12$ **35.** $-6y - 4$ **36.** $\frac{3}{2}a + 2$ **37.** $m - 3$
38. $8n + 12$ **39.** $24t - 6$ **40.** $-6a - 3$ **41.** $2d - \frac{4}{5}$
42. $-14a + 6$ **43.** $-3n - 6$ **44.** $0.5x + 3$
45. $25x + 5$ **46.** $-4t + 12$ **47.** $56b - 14$ **48.** $b + 4$
49. $18w + 6$ **50.** $-\frac{1}{4}z + 2$

3-4 ▼ Solving and Modeling Equations

1. $6x - 5$ **2.** $6a + 3$ **3.** $4y + 24$ **4.** $-5r + 3$ **5.** 1
6. −8 **7.** 15 **8.** 2 **9.** 2 **10.** −17
11. Length: 90 cm; width: 30 cm.
12. 5 lb each of peanuts and cashews

3-5 ▼ Multiplying by a Reciprocal

1. 50 **2.** 20 **3.** 45 **4.** 160 **5.** x **6.** a **7.** 30 **8.** 50 **9.** 40
10. 40 **11.** 64 **12.** 60 **13.** 45 **14.** −30 **15.** 50 **16.** 36
17. 40 **18.** −42 **19.** $.45 **20.** 6 min **21.** 4 h **22.** 25 yd

3-5 ▼ Multiplying by a Common Denominator

1. 6 **2.** 10 **3.** 12 **4.** 15 **5.** 8 **6.** 10 **7.** 50 **8.** 42 **9.** 30
10. 96 **11.** 7 **12.** 33 **13.** 90 min **14.** 180 calories

3-6 ▼ Finding the Probability of Independent Events

1. $\frac{3}{25}$ **2.** $\frac{2}{25}$ **3.** $\frac{4}{75}$ **4.** $\frac{5}{36}$ **5.** $\frac{3}{64}$ **6.** $\frac{2}{27}$ **7.** $\frac{1}{12}$

8. $\frac{5}{36}$ **9.** $\frac{5}{36}$ **10.** $\frac{5}{48}$ **11.** $\frac{1}{16}$ **12.** $\frac{3}{16}$ **13.** $\frac{1}{4}$

14. $\frac{4}{81}$ **15.** $\frac{2}{9}$

3-6 ▼ Finding the Probability of Dependent Events

1. $\frac{1}{3}$ **2.** $\frac{1}{10}$ **3.** $\frac{1}{15}$ **4.** $\frac{2}{3}$ **5.** $\frac{5}{14}$ **6.** $\frac{1}{6}$ **7.** $\frac{2}{15}$

8. $\frac{1}{10}$ **9.** $\frac{2}{15}$ **10.** $\frac{2}{15}$ **11.** $\frac{2}{15}$ **12.** $\frac{7}{30}$ **13.** $\frac{1}{380}$

14. $\frac{7}{22}$ **15.** $\frac{12}{35}$

3-6 ▽ Finding Probability Using an Equation

1. $\frac{3}{4}$ **2.** $\frac{5}{9}$ **3.** $\frac{2}{5}$ **4.** $\frac{8}{15}$ **5.** $\frac{4}{5}$ **6.** $\frac{15}{28}$ **7.** $\frac{2}{7}$ **8.** $\frac{5}{9}$
9. $\frac{1}{4}$ **10.** $\frac{3}{11}$ **11.** $\frac{1}{4}$

3-7 ▽ Solving Percent Equations

1. 0.35 **2.** 0.05 **3.** 1.25 **4.** 34% **5.** 70% **6.** 3%
7. $n = 0.10 \times 55$ **8.** $12 = 0.2n$ **9.** $n \times 40 = 25$
10. $60 = 0.3n$ **11.** $n = 0.9 \times 400$ **12.** $n \times 16 = 12$
13. $n = 0.75 \times 68$ **14.** $n \times 100 = 38$ **15.** $40 = 0.1n$
16. $x = 0.25(700)$ **17.** $1 = a(3)$ **18.** $n \times 240 = 30$
19. 15 **20.** 12 **21.** 66.7% **22.** 150 **23.** 45% **24.** 160
25. 66 **26.** 40 **27.** 64% **28.** 14.88 **29.** 37.5% **30.** $133\frac{1}{3}$%

3-7 ▽ Writing Equations to Solve Percent Problems

1. $n = 0.35 \times 40$ **2.** $20 = 0.25n$ **3.** $n \times 10 = 3$
4. $12 = 0.75n$ **5.** $n \times 30 = 20$ **6.** $n = 0.8 \times 200$
7. 63.3% **8.** 30 students **9.** $13.50 **10.** 60% **11.** 110%
12. 25%

3-7 ▽ Simple Interest

1. $b = \frac{3a}{2}$ **2.** $c = 4b - 3$ **3.** $y = \frac{0.5}{x}$ **4.** $x = 3y$
5. $m = 2n$ **6.** $h = \frac{3}{8}$ **7.** $40.15 **8.** 3.15% **9.** $844.60
10. They will both earn $15 in interest. **11.** 3.57%
12. $1250

3-8 Percent of Change

1. $3 **2.** 6 in. **3.** 50 **4.** 0.45 **5.** 0.8 cm **6.** $\frac{3}{8}$ ft
7. 25% increase **8.** 20% decrease **9.** 75% increase
10. 53.1% decrease **11.** 200% increase **12.** 10% increase
13. 21.4% increase **14.** 21.4% increase **15.** $16\frac{2}{3}$% increase
16. 20% increase **17.** 33.7% increase **18.** $12\frac{1}{2}$% decrease
19. 20% **20.** 16.7% **21.** 40% **22.** 50%

4-1 ▽ Using Properties of Equality

Answers given for **1–6** are least common denominators; answers could be any muliple of these answers. **1.** 15 **2.** 8
3. 20 **4.** 6 **5.** 36 **6.** 40 **7.** 6 **8.** 9 **9.** 5 **10.** 8 **11.** 1.2
12. 4 **13.** 10 **14.** 27 **15.** 3 **16.** 15 **17.** 4 **18.** 28
19. 6.6$\overline{6}$ **20.** 2.25 **21.** 12.6 **22.** 3 **23.** 60 **24.** 10
25. 81 **26.** 2 **27.** 5 **28.** $\frac{33}{2}$ **29.** 36 **30.** 49 **31.** $\frac{3}{2}$
32. $\frac{5}{2}$ **33.** 4 **34.** 5 **35.** 16 **36.** 155

4-1 ▽ Using Cross Products

1. 8 **2.** 7 **3.** 5.5 **4.** 2.3 **5.** 2.5 **6.** 4.5 **7.** 9 **8.** 21 **9.** 12
10. 25 **11.** 2.4 **12.** 10 **13.** −14 **14.** 16 **15.** −7.2 **16.** 6
17. 12 **18.** −100 **19.** 5 **20.** 1 **21.** 3 **22.** 4 **23.** 35
24. 14 **25.** 2 **26.** 2 **27.** 25 **28.** 10.5 **29.** 7.25 **30.** 1

4-1 ▽ Solving Percent Problems Using Proportions

1. 75 **2.** 9 **3.** 90 **4.** 24 **5.** 240 **6.** 140 **7.** 12 **8.** 50
9. 28% **10.** 50 **11.** 4.5 **12.** 23 **13.** 250 **14.** 25% **15.** 36
16. 46 **17.** 8% **18.** 30 **19.** $3000 **20.** 37.5% **21.** 400

4-2 ▽ Using Tiles to Solve Equations

1. 7 **2.** 5 **3.** 6 **4.** −8 **5.** 4 **6.** 4 **7.** 1 **8.** 2 **9.** 5 **10.** 4
11. 3 **12.** 2 **13.** $x = 4$ **14.** $a = 3$ **15.** $b = -1$

4-2 ▽ Using Properties of Equality

1. $2a + 11$ **2.** $8x$ **3.** $4y - 27$ **4.** $-3n + 7$ **5.** $4t + 5$
6. $-3q + 16$ **7.** 4 **8.** 5 **9.** 12 **10.** 3 **11.** −5 **12.** 7
13. 6 **14.** −2 **15.** 3 **16.** 7 **17.** −2 **18.** −2 **19.** $7
20. 5 h **21.** $80,000

4-2 ▽ Solving Special Types of Equations

1. $12x - 60$ **2.** $6a + 42$ **3.** $-2y - 4$ **4.** $-n + 8$
5. $2p - 10$ **6.** $10k + 100$ **7.** no solution **8.** identity
9. no solution **10.** identity **11.** no solution **12.** no
solution **13.** identity **14.** no solution **15.** identity **16.**
no solution **17.** no solution **18.** identity **19.** no
solutions **20.** identity **21.** one solution

4-3 ▽ Solving Absolute Value Equations

1. 8 **2.** 6 **3.** 6 **4.** 21 **5.** 5 **6.** 8 **7.** 3 **8.** 4 **9.** 4
10. 6 **11.** 2 **12.** 2 **13.** 7 **14.** 1 **15.** 5 **16.** 5; −5
17. 10; −10 **18.** 9; −9 **19.** 5; −5 **20.** 8; −8 **21.** 0
22. −5; −1 **23.** −5; 7 **24.** −8 **25.** 0 **26.** 1; 7 **27.** 6; −6
28. −5 or −1 **29.** −2 or 2 **30.** $1\frac{1}{3}$ or $2\frac{2}{3}$ **31.** $-4\frac{2}{3}$ or
$4\frac{2}{3}$ **32.** −17 or 13 **33.** −10 or 8

4-3 ▽ Modeling by Writing Equations

1. $x = 7$ or $x = -7$ **2.** $a - 6 = 11$ or $a - 6 = -11$
3. $n = 3$ or $n = -3$ **4.** $r + 2 = 11$ or $r + 2 = -11$
5. $t - 1 = 10$ or $t - 1 = -10$
6. $2 - e = 5$ or $2 - e = -5$ **7.** max: 13 min: −13
8. max: 5; min: −5 **9.** max: 9; min: −13 **10.** max: 9; min: 5
11. max: 6; min: −6 **12.** max: 3; min: −3 **13.** max: 17;
min: −5 **14.** max: 3; min: −3 **15.** max: 7; min: −3
16. max: $15,068; min: $14,068 **17.** max: 40°F; min: 24°F
18. $\left| x - 12(64) \right| = 2.5$; max $= 770.5$ oz; min $= 765.5$ oz.

Two-Year Algebra Handbook Answers (continued)

4-4 Transforming Formulas

1. 9 **2.** 5 **3.** 2.5 **4.** 5 **5.** -16 **6.** 3 **7.** $b = 16 - a$
8. $a = p - b - c$ **9.** $x = 10 + a$ **10.** $z = 12 - 6t$
11. $x = 4n$ **12.** $C = \pi d$ **13.** $w = \dfrac{P - 2l}{2}$

14. $x = \dfrac{3 - 5y}{2}$ **15.** $x = \dfrac{a}{n}$ **16.** $w = \dfrac{V}{lh}$

17. $y = \dfrac{5x}{3}$ **18.** $r = \dfrac{A - P}{Pt}$ **19.** $r = \dfrac{C}{2\pi}$

20. $C = \dfrac{5}{9}(F - 32)$ **21.** $\dfrac{3}{2}c - 150 = x$

4-5 ▼ Graphing and Writing Inequalities

1. false **2.** true **3.** false **4.** true **5.** true **6.** true
7. any numbers less than 1 **8.** any numbers greater than -2
9. any numbers less than or equal to 0 **10.** 0, 1, 2, 3
11. $-1, 0, 1, 2$ **12.** 1, 2, 3, 4 **13.** $-2, -3, -4, -5$
14. $-1, 0, 1, 2$ **15.** $x \geq 8$ **16.** $k < 3$ **17.** $t \leq 0$
18. $r \geq 2$ **19.** $p < 7$ **20.** $x \leq 6$ **21.** $m \geq -3$
22. $d > 8$

23.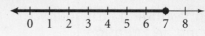

24.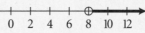

25.

26.

27.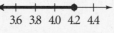

28.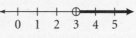

29.

30.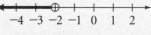

4-5 ▼ Using Addition to Solve Inequalitites

1. $-8, -3, 0, 1, 5$ **2.** $-10, -7, -5, -2, -1$ **3.** $-6, -2, 0, 2, 6$
4. $-6, -2, 1, 2, 4,$ **5.** $-8, -6, -3, -1, 0$ **6.** $-6, -5, -2, 1, 5$
7. $y < 9$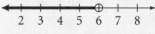

8. $x > 3$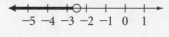

9. $a > 9$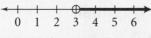

10. $n \geq 8$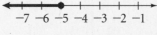

11. $k \leq -2$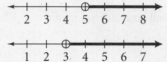

12. $t \leq 7$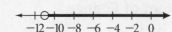

13. $a > 8$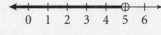

14. $r < 4$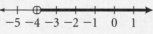

15. $m \geq 1$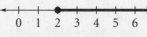

16. $x \leq 4.2$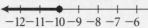

17. $b > 3$

18. $y \leq \dfrac{2}{3}$

19. $a < -2;$

20. $c > 6;$

21. $z < -\dfrac{5}{2};$

22. $x \leq -5;$

23. $t > 3;$

24. $x \geq 6;$

25. $a > 5;$

26. $x > 3;$

27. $b \geq -11;$

4-5 ▼ Using Subtraction to Solve Inequalities

1. $x > 12$ **2.** $x < 16$ **3.** $x \geq 50$ **4.** $x \geq 35$ **5.** $x \leq 75$
6. $x \leq 100$

7. $y < 5$

8. $x > -4$

9. $a > -6$

10. $n \geq 2$

11. $k \leq -10$

12. $t \le -5$

13. $a > 1$

14. $r < -2$

15. $m \ge -3$

16. $x \le 13$

17. $b > 2$

18. $y \le 3.2$

19. $37.50 or more

20. no more than six boxes;

4-6 ▼ Solving Inequalitites Using Multiplication

1. $-15 < 5$ **2.** $40 > 10$ **3.** $-30 < 0$ **4.** $-25 < -20$
5. $15 > 5$ **6.** $-15 < 0$

7. $a < 12$

8. $n \ge 15$

9. $x > -10$

10. $y \ge -20$

11. $t > 12$

12. $b \ge -12$

13. $x < 36$

14. $n \ge -77$

15. $k > -18$

16. $y < 4$

17. $a < 7$

18. $z \le 8$

4-6 ❓ Solving Inequalitites Using Division

1. $-2 < 4$ **2.** $4 > 1$ **3.** $-3 < 0$ **4.** $-5 < -2$
5. $3 > 2$ **6.** $-6 < 0$

7. $a < 2$

8. $n \ge 2$

9. $x > -2$

10. $y \ge -4$

11. $t > 5$

12. $b \ge -5$

13. 6 or more days

4-7 ▼ Solving with Variables on One Side

1. $2w + 4$ **2.** $-6x - 60$ **3.** $-5a + 80$ **4.** $5c - 45$
5. $-3t - 24$ **6.** $-r + 12$

7. $a < 4$

8. $y \ge 8$

9. $x > 6$

10. $g \ge -8$

11. $z > -21$

12. $x > -8$

13. $w \ge 31$

14. $k > -25$

15. $t \ge -8$

16. $a > 12$

17. $n \ge 16$

18. $g > 5$

19. no more than $23.75

20. no more than 18 chairs;

21. at least 15 bundles

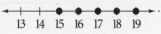

22. $k < -3$

25. $n < 10$;

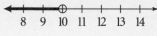

26. $e > 7$;

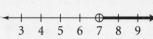

4-8 ▼ Solving Compound Inequalities Joined by *And*

1. $x < 7$ **2.** $a > 5$ **3.** $r \geq 9$ **4.** $t > -6$ **5.** $y > 9$
6. $b \leq 1$ **7.** $3 < x$ and $x < 10$ **8.** $-7 \leq d - 5$ and
$d - 5 < -2$ **9.** $6 > 2y - 7$ and $2y - 7 \geq -2$
10. $1 > x$ and $x > -1$ **11.** $t \geq 8$ and $t < 15$
12. $z > 2$ and $z < 3$ **13.** $-1 < x \leq 3$
14. $0 < x < 7$ **15.** $-8 \leq x \leq -3$
16. $-3 < x \leq 1$

17.
18.
19.
20.
21.
22.
23.
24.
25.

4-8 ▼ Solving Compound Inequalities Joined by *Or*

1. h is less than -6 or h is equal to -6 **2.** g is greater than 5 or g is equal to 5 **3.** $a + 3$ is less than -2 or $a + 3$ is equal to -2 **4.** $x - 10$ is greater than 23 or $x - 10$ is equal to 23 **5.** $5y$ is greater than 20 or $5y$ is equal to 20 **6.** $3z - 8$ is less than 16 or $3z - 8$ is equal to 16 **7.** $x < 2$ or $x \geq 6$ **8.** $x \leq -5$ or $x \geq -4$ **9.** $x < 0$ or $x > 4$

10.
11.
12.
13.
14.

4-7 ▼ Solving with Variables on Both Sides

1. $7x + 8$ **2.** $-4a - 4$ **3.** $10y - 14$ **4.** $-5n - 23$
5. $4t + 11$ **6.** $2k + 16$

7. $x \geq 11$
8. $a < 4$
9. $b \leq -10$
10. $n > -22$
11. $r \leq 33$
12. $h < -15$
13. $y \leq -2$
14. $c \leq 3$
15. $t > -5$
16. $x < 1$
17. $w \geq -33$
18. $g > -8$
19. $y \geq -3$;
20. $m > -3$;
21. $d < 1$;
22. $a > -3$;
23. $m \leq 0$;
24. $j \geq 0$;

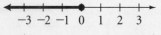

© Prentice-Hall, Inc.

15.

16.

17.

18.

4-8 ▼ Solving Absolute Values Inequalities

1. $x = 2$ or $x = -2$ **2.** $n = 15$ or $n = -15$
3. $a = 1$ or $a = -5$ **4.** $g = 3$ or $g = 9$
5. $k = 7$ or $k = -7$ **6.** $t = 0$ **7.** $|x| > 5$

8. $|x| < 2$ **9.** $|x - 7| > 3$ **10.** $|x - 8| < 5$

11.

12.

13.

14.

15.

16.

17.

18.

19.

4-9 ▼ Solving Inequalities Given a Replacement Set

1. $1, 2$ **2.** $-2, -1, 0, 1$ **3.** $-2, -1, 0, 1, 2$ **4.** $-1, 1$
5. $-1, 0, 1$ **6.** $-2, 2$
7a.

7b.

7c.

8a.

8b.

8c.

9a.

9b.

9c.

10a.

10b.

10c.

11a.

11b.

11c.

12a.

12b.

12c.

13a.

13b.

13c.

14a.

14b.

14c.

15a.

15b.

15c.

16a.

16b.

16c.

17a.

17b.
```
-3 -2 -1 0 1 2 3
```
...

17c.
```
-3 -2 -1 0 1 2 3
```
...

18a.
```
-6 -4 -2 0 2 4 6
```

18b.
```
-6 -4 -2 0 2 4 6
```

18c.
```
-6 -4 -2 0 2 4 6
```

4-9 ▼? Determining A Reasonable Answer

1. $-1, 0, 1, 2, 3, 4, 5, 6$ **2.** $3, 4, 5, \ldots$ **3.** $1, 2, 3, 4$
4. $-5, -4, -3, -2, -1$ **5.** $13x \leq 100$ **6.** $5x \geq 75$
7. $10 < 7x < 100$ **8.** 6 **9.** 2 **10.** 27 **11.** at most 5 mi

5-1 ▼ Counting Units to Find Slope

1. $\frac{4}{3}$ **2.** $\frac{5}{2}$ **3.** $\frac{3}{2}$ **4.** $\frac{2}{1}$ **5.** $\frac{3}{1}$ **6.** $\frac{1}{1}$ **7.** $\frac{2}{5}$ **8.** $\frac{3}{4}$

9. $-\frac{1}{3}$ **10.** $\frac{3}{4}$ **11.** 5 **12.** $-\frac{4}{3}$ **13.** $\frac{3}{2}$ **14.** -2

15. $-\frac{1}{3}$

5-1 ▼? Using Coordinates to Find Slope

1. -8 **2.** -3 **3.** 4 **4.** 1 **5.** -8 **6.** 14 **7.** $\frac{5}{3}$ **8.** $-\frac{1}{2}$

9. 1 **10.** $\frac{3}{4}$ **11.** $-\frac{5}{2}$ **12.** 0 **13.** undefined **14.** -1

15. $\frac{2}{5}$ **16.** $-\frac{1}{3}$ **17.** undefined **18.** 0 **19.** 2 **20.** 2

21. $\frac{5}{3}$ **22.** $\frac{4}{3}$ **23.** $-\frac{3}{2}$ **24.** $-\frac{5}{6}$ **25.** $-\frac{6}{5}$ **26.** $-\frac{6}{7}$

27. -1 **28.** -1 **29.** $-\frac{4}{5}$ **30.** -5 **31.** $\frac{4}{5}$ **32.** $\frac{1}{3}$

33. $-\frac{1}{2}$ **34.** $-\frac{3}{2}$ **35.** 0 **36.** undefined

5-1 ▼³ Graphing Lines Given a Point and Its Slope

1. positive **2.** horizontal line **3.** negative **4.** vertical line
5. negative **6.** positive

7. 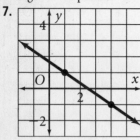 **8.**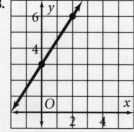

9. 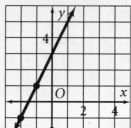 **10.**

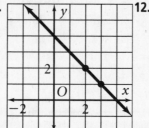

11. **12.**

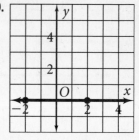

5-2 ▼ Finding Rate of Change

1. -3 **2.** 8 **3.** -11 **4.** -7 **5.** -28 **6.** -4 **7.** 20

8. $\frac{14}{1}$ (means \$14 per person) **9.** $\frac{1.3}{1}$ (means \$1.30/lb)

10. $-\frac{5}{1}$ (means 5° decrease per hour)

5-2 ▼? Using a Table

1. The length of time is the independent variable; the temperature is the dependent variable. **2.** The number of gallons of gas purchased is the independent variable; the number of miles is the dependent variable. **3.** The number of calories consumed is the dependent variable; the number of candy bars eaten is the independent variable.

4. $\frac{6}{1}$ (means \$6 per pizza) **5.** $-\frac{29}{1}$ (means 1 gal used

every 29 mi) **6.** $\frac{13}{1}$ (means \$13 per CD) **7.** $-\frac{1}{4}$ (means

\$1 per 4 games)

5-2 ▼³ Linear Functions

1. -1 **2.** $\frac{2}{3}$ **3.** $\frac{1}{3}$ **4.** $-\frac{5}{4}$ **5.** -1 **6.** 4 **7.** yes

8. no **9.** yes **10.** yes **11.** no **12.** no **13.** yes **14.** no
15. yes **16.** no **17.** yes **18.** no

5-3 ▼ Direct Variation

1. $y = x$ **2.** $y = \frac{1}{3}x$ **3.** $y = 5x - 7$ **4.** $y = 2x$

5. $y = -4x + 9$ **6.** $y = \frac{3}{5}x$ **7.** $y = 2x + 7$

8. $y = \frac{3}{4}x - \frac{9}{4}$ **9.** $y = \frac{1}{4}x$ **10.** yes; $-\frac{2}{5}$ **11.** no

12. yes; $\frac{7}{6}$ 13. no 14. yes; $-\frac{7}{4}$ 15. no 16. no

17. yes; $\frac{2}{9}$ 18. yes; -1 19. yes; $\frac{1}{4}$ 20. no 21. yes; 6

22. no 23. no 24. yes; $\frac{1}{8}$ 25. no 26. yes; $\frac{1}{7}$ 27. no

28. $\frac{3}{2}$ 29. -2 30. $\frac{4}{3}$ 31. $\frac{2}{5}$ 32. -4 33. 4 34. 3

35. 5 36. $\frac{3}{2}$ 37. 2 38. $\frac{2}{3}$ 39. $-\frac{3}{4}$

5-3 ▼ Using the Constant of Variation to Write Equations

1. 6 2. $\frac{7}{3}$ 3. 0.35 4. 0.35 5. 6 6. $\frac{7}{3}$ 7. yes; $y = \frac{1}{3}x$

8. no 9. yes; $y = -\frac{7}{2}x$ 10. $y = 2.54x$ 11. $y = 29.6x$

12. $560

5-3 ▼ Using Proportions

1. 8 2. 35 3. 81 4. 52 5. 12 6. 48 7. 21 8. 9 9. -25

10. 15 11. 3 12. 14 13. 5 14. -14 15. $\frac{5}{2}$ 16. $6\frac{1}{4}$

17. 0.55 L 18. 164 mi 19. 20.8 lb 20a. $\frac{32}{7}$ 20b. $\frac{5}{12}$

20c. $\frac{3}{4}$

5-4 ▼ Defining Slope-Intercept Form

1. up 3, right 1; down 3, left 1   2. down 4, right 1; up 4, left 1
3. up 0, right 1; up 0, left 1   4. up 1, left 2; down 1, right 2
5. up 2, right 3; down 2, left 3   6. up 5, left 2; down 5, right 2
7. up 9, right 1; down 9, left 1   8. up 4, right 7; down 4, left 7
9. up 2, left 1; down 2, right 1   10. up 6, right 1; down 6, left 1
11. $m = 2; b = -3$ 12. $m = -4; b = 7$
13. $m = -2; b = -8$ 14. $m = 1; b = 3$
15. $m = 0; b = 9$ 16. $m = 3; b = 2$
17. $m = -1; b = 5$ 18. $m = 7; b = 0$
19. $m = 6; b = 5$ 20. $m = -9; b = 2$
21. $m = 5; b = -13$ 22. $m = -\frac{1}{2}; b = -7$

23. $m = 0; b = 14$ 24. $m = \frac{3}{4}; b = 6$

25. $m = -11; b = 0$ 26. $-2; 0$ 27. $-5; 0$ 28. 3; 6

29. $y = -3x + 2$;

30. $y = 4x - 1$;

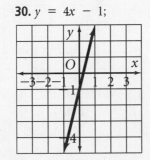

31. $y = -3x - 1$;

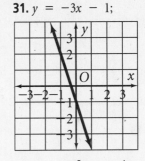

32. $y = 3x + 2$;

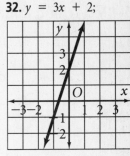

33. $y = -\frac{2x}{3} + \frac{4}{3}$;

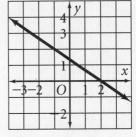

34. $y = \frac{3x}{2} - 4$;

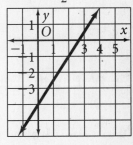
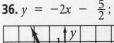

35. $y = x + 2$;

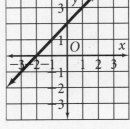

36. $y = -2x - \frac{5}{2}$;

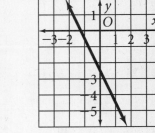

37. $y = \frac{3}{2}x + \frac{9}{4}$;

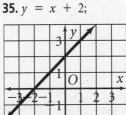

38. $y = \frac{8}{3}x + \frac{7}{3}$;

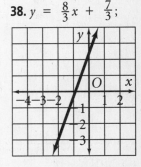

39. $y = -\frac{1}{3}x - \frac{4}{3}$;

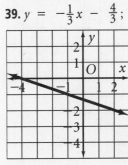

40. $y = -\frac{1}{3}x - 2$;

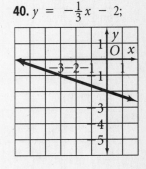

5-4 ▼ Writing Equations

1. $A = \frac{bh}{2}$ 2. $C = 2\pi r$ 3. $P = 2l + 2w$ 4. $y = 2x$

5. $y = 2$ 6. $y = 2x + 2$ 7. $y = \frac{1}{2}x - 2$

8. $y = -\frac{3}{2}x + 1$ 9. $y = -2x - 2$ 10. $y = \frac{1}{4}x + 2$

11. $y = \frac{2}{3}x - 2$ 12. $y = 5$ 13. $y = -3x$

14. $y = -\frac{1}{2}x - 3$ 15. $y = -\frac{1}{3}x + 1$

16. $y = x + 5$ 17. $y = 6x - 8$ 18. $y = \frac{2}{3}x + \frac{1}{3}$

19. $y = -3x + 2$ 20. $y = -\frac{1}{2}x - 4$ 21. $y = 5x$

5-5 Writing the Equation of a Line

1. 21 2. −4 3. −7 4. 11 5. 4 6. −8 7. −16 8. 5 9. 3

10. $m = \frac{1}{9}$ 11. $m = \frac{1}{11}$ 12. $m = \frac{1}{6}$ 13. $m = -2$

14. $m = \frac{9}{5}$ 15. $m = -1$ 16. $m = 2$ 17. $m = \frac{10}{9}$

18. $m = 0$ 19. $y = \frac{11}{14}x + \frac{5}{2}$ 20. $y = -4x$

21. $y = \frac{1}{2}x + 7$ 22. $y = \frac{4}{5}x - \frac{51}{5}$

23. $y = -x + 5$ 24. $y = x - 3$ 25. $y = \frac{3}{2}x - 9$

26. $y = \frac{1}{12}x - \frac{17}{12}$ 27. $y = -\frac{7}{9}x + 5$

28. $y = 2x - 1$ 29. $y = -\frac{4}{3}x$ 30. $y = 1$

31. $y = 2x$ 32. $y = -\frac{1}{5}x + \frac{23}{5}$ 33. $y = \frac{1}{2}x - \frac{7}{2}$

34. $y = x - 1$ 35. $y = -\frac{2}{3}x + \frac{13}{3}$ 36. $y = \frac{1}{3}x$

5-6 ▼ Trend Line

1.
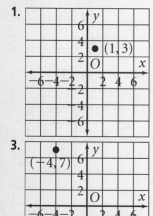

2.
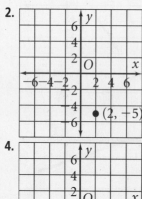

3.
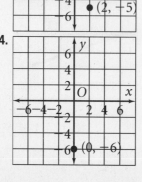

4.

5.
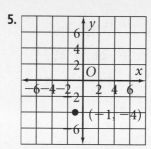

6.
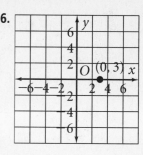

7. Answers may vary. Sample: $y = -x + 2$

8. Answers may vary. Sample: $y = x - 3$

9. Answers may vary. Sample: $y = \frac{5}{7}x + 5$

10. Answers may vary. Sample: $y = -\frac{3}{4}x + 3$

11. Answers may vary. Sample: $y = \frac{5}{3}x + 5$

12. Answers may vary. Sample: $y = \frac{4}{3}x - 4$

5-6 ▼ Line of Best Fit

1.
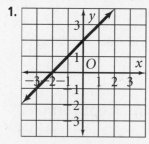

2.

3.
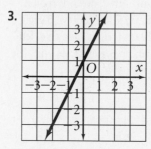

4.
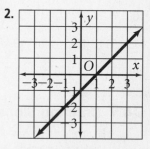

5.
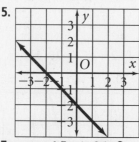

6.
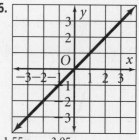

7. $y = -1.7x + 8.1$ 8. $y = 1.55x - 3.95$

9. $y = 1.57x + 2.25$ 10. $y = -x + 8.3$

11. $y = -1.78x + 15.55$ 12. $y = 3x - 3.2$

13. $y = -2.1x + 9.1$ 14. $y = -0.2x + 34.8$

5-7 ▼ Graphing Equations

1. $x = 2$ **2.** $x = -2$ **3.** $x = -6$ **4.** $x = \frac{5}{3}$

5. $x = 4$ **6.** $x = \frac{7}{13}$ **7.** $x = -\frac{3}{2}$ **8.** $x = \frac{3}{4}$

9. $x = -2$

10.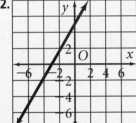

11.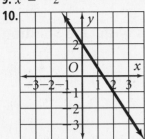

12.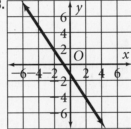

13.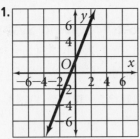

14.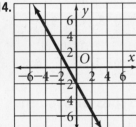

15.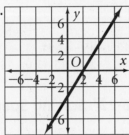

16.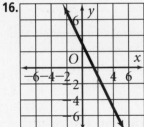

17.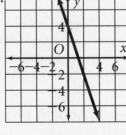

18.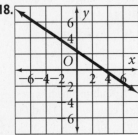

19.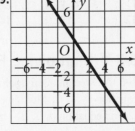

20.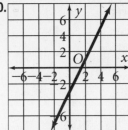

21.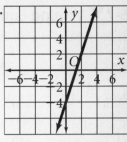

22.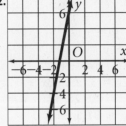

23.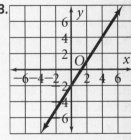

24.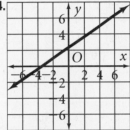

25.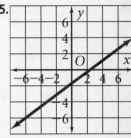

26.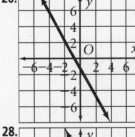

27.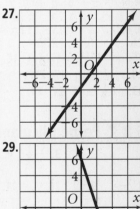

28.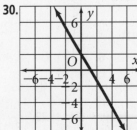

29.

30.

5-7 ▼ Writing Equations

1. $2x = 12, x = 6$ **2.** $7x = -45, x = -\frac{45}{7}$

3. $5x = -48, x = -\frac{48}{5}$ **4.** $2x = -40, x = -20$

5. $8x = 21, x = \frac{21}{8}$ **6.** $7x = 4, x = \frac{4}{7}$ **7.** $x = 1$

8. $x = 18$ **9.** $x = 12$ **10.** $3x + 7y = 43$
11. $-4x + y = 1$ **12.** $-6x + 5y = 22$
13. $2x + y = -12$ **14.** $-x + 9y = -28$
15. $8x + 3y = -31$ **16.** $-5x + y = 1$
17. $3x + 4y = -5$ **18.** $4x + y = -17$
19. $-x + 2y = -15$ **20.** $8x + y = 56$
21. $-6x + 5y = 18$ **22.** $-2x + y = -1$
23. $-x + y = 0$ **24.** $-x + 2y = 3$ **25.** $-3x + y = 4$
26. $-x + y = 3$ **27.** $2x + 3y = 10$ **28.** $3x + y = 17$
29. $x + y = 0$ **30.** $x + 3y = 9$ **31.** $x + y = 3$
32. $-2x + 3y = 0$ **33.** $y = 3$ **34.** $-x + y = 0$
35. $x + y = 4$ **36.** $x + 2y = 4$ **37.** $-3x + y = 7$
38. $-2x + 7y = 10$ **39.** $3x + 8y = 0$
40. $-5x + y = 5$ **41.** $-7x + 2y = 4$ **42.** $y = -5$
43. $2x + y = 12$ **44.** $8x + 3y = 0$ **45.** $y = 3$

5-8 ▼ Parallel Lines

1. $m = 7$ **2.** $m = $ undefined **3.** $m = \frac{2}{5}$ **4.** $m = \frac{5}{3}$

5. $m = -4$ **6.** $m = 0$ **7.** $m = \frac{4}{3}$ **8.** $m = \frac{1}{2}$

9. $m = \frac{5}{7}$ **10.** yes **11.** yes **12.** no **13.** no **14.** yes

15. no **16.** no **17.** yes **18.** yes **19.** $y = -x + 10$
20. $y = 6x + 17$ **21.** $y = -5x - 49$ **22.** $y = 2x + 1$
23. $y = 3x + 21$ **24.** $y = -4x + 32$

5-8 ▼ Perpendicular Lines

1. $-\frac{3}{7}$ **2.** $\frac{5}{2}$ **3.** $-\frac{1}{4}$ **4.** $\frac{1}{6}$ **5.** $\frac{4}{9}$ **6.** $-\frac{8}{5}$ **7.** -1

8. $-\frac{3}{4}$ **9.** $\frac{8}{3}$ **10.** yes **11.** no **12.** no **13.** yes **14.** yes

15. no **16.** $y = \frac{5}{12}x + 13$ **17.** $y = -\frac{7}{6}x + \frac{14}{3}$

18. $y = -x + 7$ **19.** $y = \frac{10}{3}x - \frac{17}{3}$

20. $y = -\frac{1}{2}x - 2$ **21.** $y = -\frac{4}{5}x - 1$

5-9 Using the x-intercept

1. 2 **2.** $\frac{25}{8}$ **3.** 15 **4.** -14 **5.** -27 **6.** $-\frac{14}{3}$ **7.** 6 **8.** 9

9. -6 **10.** 11 **11.** 6 **12.** $\frac{4}{15}$ **13.** $-\frac{3}{5}$ **14.** $-\frac{20}{9}$ **15.**

$\frac{7}{2}$ **16.** 9 **17.** 1 **18.** -5 **19.** -2 **20.** 1 **21.** 4 **22.** $\frac{37}{3}$

23. $-\frac{10}{3}$ **24.** 27 **25.** 2 **26.** 8 **27.** $-\frac{32}{5}$ **28.** $\frac{5}{8}$

29. -10 **30.** $\frac{36}{5}$ **31.** $-\frac{16}{3}$ **32.** 6 **33.** 2 **34.** 6 **35.** 28

36. -4

6-1 ▼ Solving Systems with One Solution

1. 10; 4 **2.** 8; 0 **3.** 4.25; 3.75 **4.** 2; -1 **5.** -23; -13

6. $\frac{3}{5}$; $\frac{1}{2}$ **7.** $(2, 2)$ **8.** $(-2, 1)$ **9.** $(4, -3)$ **10.** $(-5, -5)$

11. $(3, 0)$ **12.** $(0, -2)$ **13.** $(-3, 2)$ **14.** $(1, -4)$ **15.** $(4, 0)$

6-1 ▼ Solving Special Types of Systems

1. $y = 3x - 6$ **2.** $y = 5x$ **3.** $y = \frac{1}{3}x + 5$

4. $y = 9x + 27$ **5.** $y = 3x + 13$ **6.** $y = \frac{1}{6}x - \frac{5}{6}$

7. no solution **8.** infinitely many solutions **9.** no solution
10. no solution **11.** infinitely many solutions **12.** no solution

6-2 ▼ Solving Systems with One Solution

1. $z = -2$ **2.** $p = -3$ **3.** $b = 4$ **4.** $a = 1.5$

5. $d = -18$ **6.** $f = \frac{4}{5}$ **7.** $(3, 4)$ **8.** $(4, 0)$ **9.** $(-3, 2)$

10. $(-4, -4)$ **11.** $(1, -3)$ **12.** $(-1, 1)$ **13.** $(2, 2)$
14. $(-2, 1)$ **15.** $(3, 0)$ **16.** $(-2, -6)$ **17.** $(2, 4)$ **18.** $(0, -2)$

6-2 ▼ Solving Special Types of Systems

1. $n = 5m - 7$ **2.** $x = \frac{1}{2}y$ **3.** $x = 2y + 10$

4. $b = 2a - 3$ **5.** $f = 6g - 8$ **6.** $c = -\frac{3}{4}d + \frac{1}{2}$

7. infinitely many solutions **8.** no solution **9.** infinitely
any solutions **10.** no solution **11.** no solution
12. infinitely many solutions **13.** infinitely many solutions
14. no solution **15.** infinitely many solutions

6-3 ▼ Adding or Subtracting Equations

1. add **2.** subtract **3.** add **4.** add **5.** subtract **6.** add
7. $(3, -3)$ **8.** $(1, 1)$ **9.** $(1, 2)$ **10.** $(3, 4)$ **11.** $(4, 7)$
12. $(-5, 8)$ **13.** $(15, 1)$ **14.** $(3, 6)$ **15.** $(1, -1)$

6-3 ▼ Multiplying First

1–6. Answers may vary. Samples: **1.** Multiply the first
equation by 4. **2.** Multiply the first equation by 3 and the
second equation by 2. **3.** Multiply the first equation by 5.
4. Multiply the first equation by either 4 or 7. **5.** Multiply
the first equation by 8. **6.** Multiply the second equation by 3.
7. $(4, 6)$ **8.** $(-3, 4)$ **9.** $(20, -15)$ **10.** $(-3, -8)$

11. $(3.5, 2.5)$ **12.** $(10, -2)$ **13.** $(-3, 2)$ **14.** $(1, -3)$
15. $(4, 0)$ **16.** $(-5, -5)$ **17.** $(3, 0)$ **18.** $(0, -2)$

6-4 Writing Systems

1. substitution **2.** elimination **3.** substitution **4.** about 80 wooden coasters, about 135 steel coasters **5.** 36 CDs
6. 100 visitors

6-5 Linear Inequalities

1. true **2.** true **3.** false **4.** true **5.** false **6.** true

7. 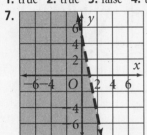 **8.**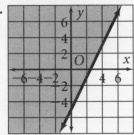

9. 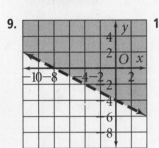 **10.**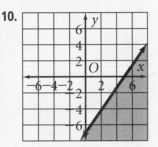

11. 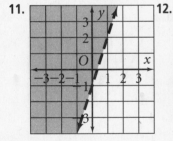 **12.**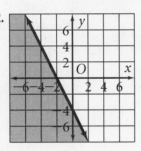

13.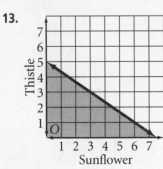

14.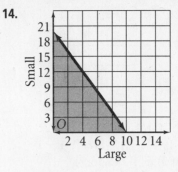

6-6 Systems of Linear Inequalities

1. $(-3, -2)$ **2.** $(2, 8)$ **3.** $(-1, -4)$

4. 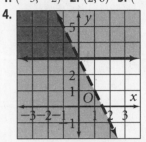 **5.**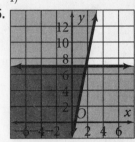

6. 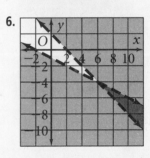 **7.**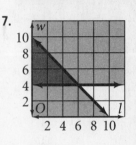

6-7 Concepts of Linear Programming

1. $(5, 3)$ **2.** $(0, 9)$ **3.** $(6, 2)$ **4.** maximum: 29 minimum: 7
5. maximum: 32 minimum: 16 **6.** maximum: 15 minimum: 7
7. 5 t of type A, 10 t of type B

6-8 Systems with Nonlinear Equations

1–6. Answers may vary.

1.

−2	6
−1	3
0	2
1	3
2	6

2.

−2	5
−1	−1
0	−3
1	−1
2	5

3.

−2	−1
−1	0
0	1
1	0
2	−1

4.

−2	5
−1	2
0	−1
1	2
2	5

Two-Year Algebra Handbook Answers (continued)

5.

−2	−8
−1	−2
0	0
1	−2
2	−8

6.

−2	4
−1	0
0	−4
1	0
2	4

7. $(−1, 3)$ $(2, 6)$ **8.** $(−2, 7)$ $(1, 4)$ **9.** $(3, 13)$ $(0, −5)$
10. $(2, 2)$ $(−2, 2)$ **11.** $(2, −6)$ $(−1, 3)$ **12.** $(−4, −5)$ $(1, 1)$
13. $(−3, 0)$ $(2, 5)$ **14.** $(0, 3)$ $(5, 4)$ **15.** $(3, 4)$ $(−1, 0)$

7-1 ▼ Quadratic Functions

1. $y = 2x^2 + 3x + 24$ **2.** $y = 2x^2 + 2x − 3$
3. $y = 2x^2 − 3x + 1$ **4.** $y = −3x^2 + 3x + 3$
5. $y = −9x^2 − 12x$ **6.** $y = −8x^2 + 4$
7. $y = 2x^2 − 18x + 3; a = 2; b = −18; c = 3$

8. $y = \frac{1}{4}x^2 − 3x − 4; a = \frac{1}{4}; b = −3; c = −4$

9. $y = −x^2 + 3x + 5; a = −1; b = 3; c = 5$

10. $y = −0.75x^2 + 4.5x − 1.75; a = −0.75;$
$b = 4.5; c = −1.75$
11. $3x^2 − 2x + 11 = 0; a = 3; b = −2; c = 11$
12. $5x^2 + 3x − 10 = 0; a = 5; b = 3; c = −10$
13. $3x^2 − x + 2 = 0; a = 3; b = −1; c = 2$
14. $x^2 − 4x − 9 = 0; a = 1; b = −4; c = −9$
15. $x^2 − 8x − 8 = 0; a = 1; b = −8; c = −8$
16. $2x^2 − 800 = 0; a = 2; b = 0; c = −800$
17. $16x^2 − 6x − 4 = 0; a = 16; b = −6; c = −4$
18. $3x^2 + 2x − 8 = 0; a = 3; b = 2; c = −8$
19. $x^2 + 4x + 4 = 0; a = 1; b = 4; c = 4$
20. $4x^2 + 11x + 7 = 0; a = 4; b = 11; c = 7$
21. $2x^2 − 6x = 0; a = 2; b = −6; c = 0$
22. $3x^2 − 8x − 4 = 0; a = 3; b = −8; c = −4$
23. $a = 1; b = −1; c = −2$
24. $a = 1; b = −6; c = 9$
25. $4x^2 − x + 80 = 0; a = 4; b = 1; c = 80$
26. $z^2 − 4z + 3 = 0; a = 1; b = −4; c = 3$
27. $6m^2 − 72 = 0; a = 6; b = 0; c = −72$
28. $a = 1; b = 8; c = −20$ **29.** $a = 4; b = 3; c = −1$
30. $a = 2; b = 5; c = 3$ **31.** $3x^2 −6x − 2; a = 3;$
$b = −6; c = −2$ **32.** $x^2 − 4x + 3a = 1; b = −4;$
$c = 3$ **33.** $x^2 + 5x + 6 = 0; a = 1; b = 5; c = 6$
34. $a = 2; b = −5; c = −12$ **35.** $4x^2 − 12x + 9 = 0;$
$a = 4; b = −12; c = 9$ **36.** $a = 6; b = 7; c = −5$
37. $2x^2 − 2x − 5 = 0; a = 2; b = −2; c = −5$
38. $x^2 − 3x + 2 = 0; a = 1; b = −3; c = 2$
39. $3x^2 − 4x − 7 = 0; a = 3; b = −4; c = −7$
40. $3x^2 − 5x − 6 = 0; a = 3; b = −5; c = −6$

7-1 ▼ The role of a

1. $y = x^2 + 5x + 2; a = 1$
2. $y = 2x^2 + 2x − 3; a = 2$
3. $y = 3x^2 − 4x + 1; a = 3$
4. $y = −x^2 − x + 3; a = −1$
5. $y = −6x^2 − 12x + 3; a = −6$
6. $y = −3x^2 + 4; a = −3$
7–10: equations are listed in order of widest to narrowest.
7. $y = x^2$ opens upward, $y = 2x^2$ opens upward, $y = 4x^2$
opens upward

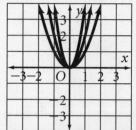

8. $y = \frac{1}{4}x^2$ opens upward, $y = x^2$ opens upward,
$y = 3x^2$ opens upward

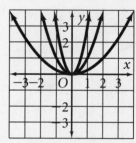

9. $y = \frac{1}{3}x^2$ opens upward, $2y = 4x^2$ opens upward,
$y = 3x^2$ opens upward

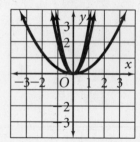

10. $y = −0.75x^2$ opens downward, $y = x^2$ opens upward,
$y = −4x^2$ opens downward

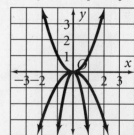

11. $y = 0.5x^2$ opens upward, $y = x^2$ opens upward, $y = -3x^2$ opens downward

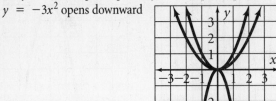

12. $y = 3x^2$ opens upward, $y = -4x^2$ opens downward, $y = 9x^2$ opens upward

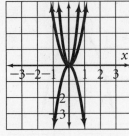

13. $y = \frac{1}{3}x^2$ opens upward, $y = \frac{2}{3}x^2$ opens upward, $y = \frac{3}{2}x^2$ opens upward,

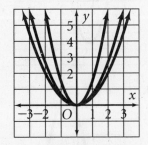

14. $y = \frac{1}{3}x^2$ opens upward, $y = 3x^2$ opens upward, $y = -3x^2$ opens downward

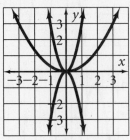

15. $y = x^2$ opens upward, $y = 1.5x^2$ opens upward, $y = 4x^2$ opens upward

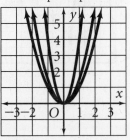

16. $y = -2x^2$ opens downward, $y = -2.5x^2$ opens downward, $y = -4x^2$ opens downward

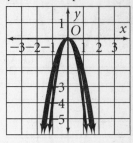

7-2 Graphing Simple Quadratic Functions

1. upward, above **2.** upward, below **3.** downward, below
4. downward, above **5.** downward, above **6.** upward, above

7.

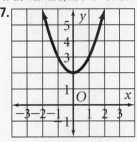

8.

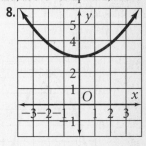

9.

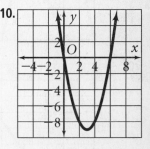

10.

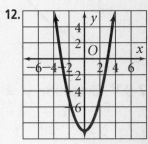

11.

12.

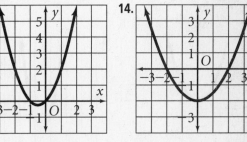

13.

14.

Two-Year Algebra Handbook Answers (continued)

15.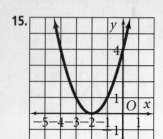

16.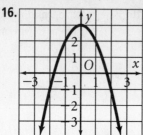

9.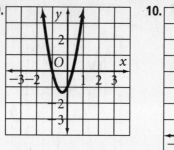

10.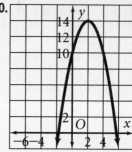

17.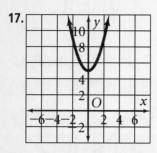

18.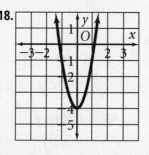

11.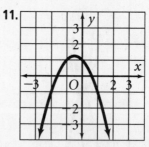

12.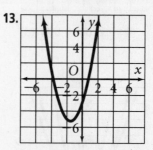

19.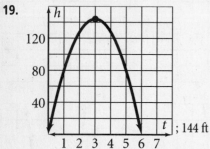

; 144 ft

13.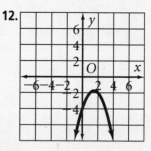

14.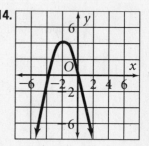

20.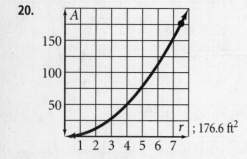

; 176.6 ft²

15.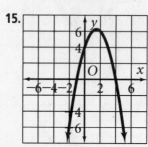

7-3 ▼ Graphing $y = ax^2 + bc + c$

1. $y = 2x^2 + 4x + 2$; $x = -1$ **2.** $y = 2x^2 + 8x - 32$;
$x = -2$ **3.** $y = 2x^2 - 8x + 1$; $x = 2$
4. $y = -12x^2 + 6x + 6$; $x = 0.25$
5. $y = -6x^2 - 12x$; $x = -1$
6. $y = -8x^2 + 24$; $x = 0$

7.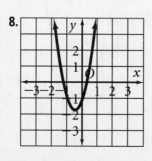

8. [graph]

16.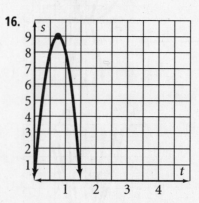

The maximum
height reached is 9 ft
(after 0.75 s).

7-3 ▽ Quadratic Inequalities

1. yes **2.** no **3.** yes

4.

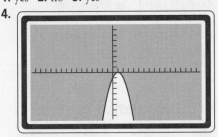 ; dashed

5.

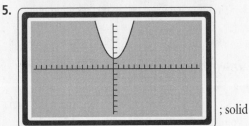 ; solid

6.

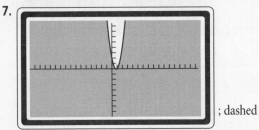 ; solid

7.

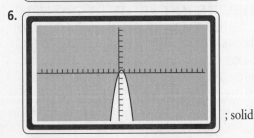 ; dashed

8.

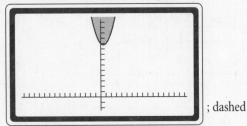 ; dashed

9.

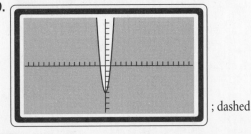 ; dashed

10.

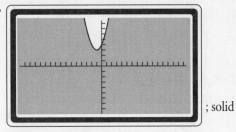 ; solid

11.

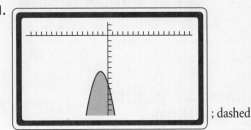 ; dashed

12.

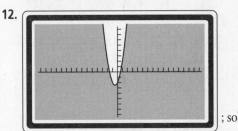 ; solid

13.

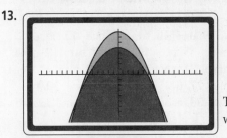

The car will fit.

14.

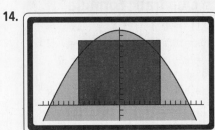

The locomotive will not fit through the tunnel, since some points that represent the locomotive go beyond the shaded area.

7-4 ▽ Finding Square Roots

1. 9 **2.** $\frac{9}{100}$ **3.** 36 **4.** 100 **5.** $\frac{81}{25}$ **6.** $\frac{25}{49}$ **7.** -3

8. undefined **9.** ±4 **10.** $\frac{8}{9}$ **11.** 2.65 **12.** 11 **13.** 0.9

14. undefined **15.** 1.1 **16.** ±2 **17.** undefined **18.** -5

19. -3.74 **20.** 6 **21.** ±0.5 **22.** $-\frac{1}{5}$ **23.** ±0.3

24. undefined **25.** -7 **26.** 0.8 **27.** ±9 **28.** -1.5 **29.** $\frac{5}{6}$

30. ±13 **31.** 10.49 **32.** 16 **33.** undefined

7-4 ▼ Estimating and Using Square Roots

1. 2.24 **2.** 4.69 **3.** 4.12 **4.** 10.95 **5.** -9.95 **6.** -11.09
7. $16, 25$ **8.** $36, 49$ **9.** $81, 100$ **10.** 9.11 **11.** $121, 144$
12. $25, 36$ **13.** $64, 81$ **14.** 671
15. r is between 4 and 5; $r = 4.12$
16. $l = 24$ in.; $w = 8$ in.

7-5 ▼ Using Square Roots to Solve Equations

1. 6 **2.** 8 **3.** 11 **4.** 13 **5.** 4 **6.** 9 **7.** ±3 **8.** ±2.83
9. ±3 **10.** ±5 **11.** ±6 **12.** ±2.65 **13.** $\pm\frac{1}{2}$
14. ±5 **15.** ±9 **16.** ±6 **17.** ±2 **18.** ±16 **19.** ±1.73
20. ±8 **21.** ±2 **22.** ±9 **23.** ±6 **24.** ±7 **25.** ±2
26. ±5.66 **27.** ±6 **28.** $l = 8$ ft; $w = 4$ ft **29.** $-8, 2$
30. $20\sqrt{5}$ cm **31.** 30 cm **32.** $9, 4$

7-5 ▼ Finding the Number of Solutions

1. ±6 **2.** ±8 **3.** ±10 **4.** $\pm2\sqrt{5}$ **5.** $\pm5\sqrt{5}$ **6.** $\pm6\sqrt{2}$
7. none **8.** two **9.** one **10.** two **11.** one **12.** none
13. two **14.** none **15.** two **16.** two **17.** two **18.** none
19. two **20.** none **21.** none **22.** one **23.** two **24.** none
25. two **26.** one **27.** two **28.** two **29.** none **30.** two
31. one **32.** two **33.** none **34.** two **35.** none **36.** two

7-6 Using the Quadratic Formula

1. $-3x^2 + 2x - 11 = 0, a = 3, b = -2, c = 11$
2. $5x^2 + 3x - 10 = 0, a = 5, b = 3, c = -10$
3. $3x^2 - x + 2 = 0, a = 3, b = -1, c = 2$
4. $-x^2 + 4x + 9 = 0, a = -1, b = 4, c = 9$
5. $x^2 - 8x - 8 = 0, a = 1, b = -8, c = -8$
6. $2x^2 - 800 = 0, a = 2, b = 0, c = -800$

7. $-1, 2$ **8.** 3 **9.** $\frac{1}{3}, 1$ **10.** $-3 \pm 2\sqrt{3}$ **11.** $3, -1$

12. $2, -\frac{1}{2}$ **13.** -2 **14.** $1, 6$ **15.** $-0.35, 0.72$

16. $-1, -1.75$ **17.** $3, 0$ **18.** -2 **19.** $3.1, -0.43$

20. $-2, \frac{4}{3}$ **21.** $3.31, -1.81$ **22.** no solution

23. no solution **24.** $2, 1$

7-7 Using the Discriminant

1. 0 **2.** -7 **3.** 8 **4.** 64 **5.** 9 **6.** 0 **7.** one solution
8. two solutions **9.** two solutions **10.** no solution
11. two solutions **12.** two solutions **13.** two solutions

14. two solutions **15.** two solutions **16.** two solutions
17. two solutions **18.** one solution **19.** two solutions
20. no solution **21.** no solution **22.** two solutions
23. two solutions **24.** two solutions **25.** one solution
26. one solution **27.** no solution **28.** no solution
29. two solutions **30.** no solution **31.** two solutions
32. one solution **33.** one solution

8-1 ▼ Exploring Exponential Patterns

1. 40 **2.** 90 **3.** $1,024$ **4.** 50 **5.** 216 **6.** 243 **7.** $3,200$ students **8.** 32 million vehicles **9.** 2,700,000 microchips **10.** $8000

8-1 ▼ Evaluating Exponential Functions

1. $3(4^2)$ **2.** $8(3^3)$ **3.** 2^5 **4.** 5^3 **5.** $10(6^3)$ **6.** 3^3 **7.** $162, 486, 1458$ **8.** $48, 192$ **9.** $5, 25, 125, 625$ **10.** $196, 1372, 9604$
11. $512; 32,768$ **12.** $320, 640, 1280$ **13.** $48; 768; 12,288$
14. $81, 729, 6561$ **15.** $16, 64, 1024$ **16.** $392, 2744$ **17.** $320, 5120$ **18.** $125; 3125; 78,125$ **19.** $20,000; 200,000; 2,000,000$
20. $108, 648, 3888$ **21.** $81, 6561$ **22.** $15, 45, 405$ **23.** $1, 1$
24. $1000, 10000$ **25.** $28, 112$ **26.** $121, 1331$ **27.** $36, 108$
28. $448, 28672$ **29.** $100, 500, 2500$ **30.** $54, 486, 4374$
31. $98, 4802, 33614$ **32.** $375, 9375$ **33.** $10, 10, 10$ **34.** $32, 512, 8192$ **35.** $18, 72, 288$ **36.** $147, 7203$ **37.** $30, 180, 1080$
38. $49, 343, 2401$ **39.** $32, 128, 256$ **40.** $36, 108, 324$

8-1 ▼ Graphing Exponential Functions

1. 125 **2.** 243 **3.** 81 **4.** 3.0625 **5.** 16 **6.** 15.625
7.

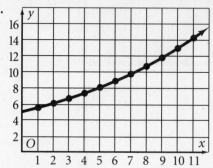

8.

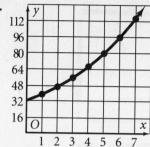

9.

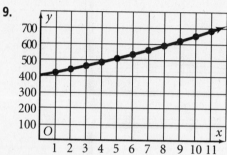

10.

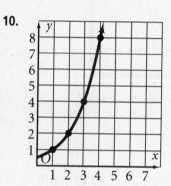

11.

12.

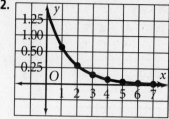

13.

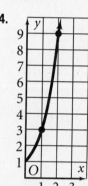

14.

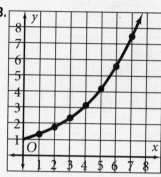

15.

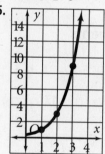

8-2 ▼ Modeling Exponential Growth

1. 1.27   **2.** 162.98   **3.** 156.82   **4.** 46.89   **5.** 2.81   **6.** 43.66
7. $42,213.01   **8.** 454 members

8-2 ▼ Finding Compound Interest

1. 1.80   **2.** 2.56   **3.** 1.99   **4.** 1.22   **5.** 2.15   **6.** 8.27
7. $234.52   **8.** $960.06   **9.** $5.18

8-3 Exponential Decay

1. 25.92 **2.** 1850.29 **3.** 44.67 **4.** 2.14 **5.** 157.27
6. 18.28

7. 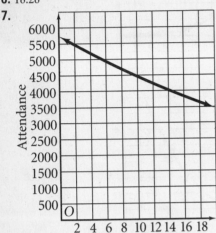 ; 3933 people

Attendance / Years of Decline

8. 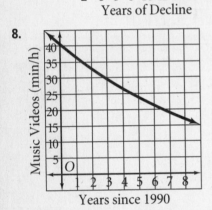 ; 24.03 min/h

Music Videos (min/h) / Years since 1990

8-4 ▽ Using Zero and Negtive Integers as Exponents

1. $\frac{1}{128}$ **2.** $\frac{1}{256}$ **3.** $\frac{1}{729}$ **4.** $\frac{1}{1000}$ **5.** $\frac{1}{25}$ **6.** 1 **7.** $\frac{1}{81}$

8. $\frac{1}{16}$ **9.** 1 **10.** $\frac{1}{216}$ **11.** $\frac{1}{64}$ **12.** $\frac{1}{1331}$ **13.** $\frac{1}{1024}$

14. 1 **15.** $\frac{1}{625}$ **16.** $-\frac{1}{27}$ **17.** 1 **18.** $\frac{1}{144}$ **19.** $\frac{1}{6561}$

20. $\frac{1}{169}$ **21.** $\frac{1}{512}$ **22.** $-\frac{1}{1024}$ **23.** $\frac{1}{117,649}$

24. $-\frac{1}{512}$ **25.** $\frac{1}{1024}$ **26.** $\frac{1}{196}$ **27.** $\frac{1}{100,000}$ **28.** $\frac{1}{256}$

29. 1 **30.** $\frac{1}{256}$ **31.** $\frac{1}{177,147}$ **32.** $-\frac{1}{125}$ **33.** 1

8-4 ▽ Relating the Properties to Exponential Functions

1. 25 **2.** $\frac{1}{27}$ **3.** $\frac{1}{216}$ **4.** 256 **5.** $\frac{1}{343}$ **6.** $\frac{1}{1331}$

7.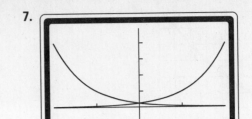

8.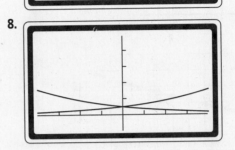

9.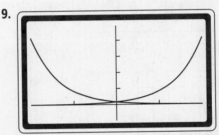

10.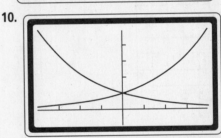

11.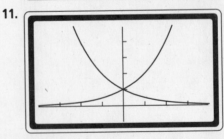

12.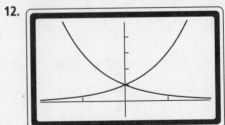

Two-Year Algebra Handbook Answers (continued)

8-5 ▼ Writing Numbers in Scientific Notation

1. no 2. yes; 3.348; 18 3. no 4. yes; 9.9325; 2
5. yes; 7.25; -3 6. no

7.

Planet	Approximate Mass (kg), standard notation
Mercury	318,810,000,000,000,000,000,000
Mars	641,800,000,000,000,000,000,000
Pluto	1,080,000,000,000,000,000,000,000
Venus	4,883,000,000,000,000,000,000,000
Earth	5,979,000,000,000,000,000,000,000
Uranus	86,820,000,000,000,000,000,000,000
Neptune	102,700,000,000,000,000,000,000,000
Saturn	568,400,000,000,000,000,000,000,000
Jupiter	1,901,000,000,000,000,000,000,000,000

8-5 ▼ Calculating With Scientific Notation

1. 10^3 2. 10^{19} 3. 10^{15} 4. 10^9 5. 10^{33} 6. 10^{30}
7. 8×10^7 8. 6.3×10^{13} 9. 1.6×10^9 10. 2×10^8
11. 3×10^{12} 12. 4.2105×10^4 13. 1.44×10^9
14. 5.6×10^{14} 15. 3.5×10^9 16. 2.1×10^{38}
17. 7×10^{12} 18. 3.33×10^5 19. 1×10^9
20. 4.35×10^{27} 21. 5×10^9 22. 8.949×10^6
23. 6.6×10^{19} 24. 3×10^5 25. 1.2825×10^6
26. 6×10^{17} 27. 8.25×10^{63} 28. 1.901×10^{27} kg
29. 1.991×10^{30} kg 30. $9.506\,606 \times 10$ Earth Masses

8-6 ▼ Multiplying Powers

1. $(x-7)^5$ 2. y^8 3. 9^4v^3 4. $2^2j^2k^5$ 5. $4^3s^3t^2$
6. $z^5(x+3)^2$ 7. $8c^7d^3$ 8. $112a^{20}b$ 9. $50e^{13}f^6$
10. $42g^2h^{10}$ 11. $245j^6k^6$ 12. $63l^8m^{11}$ 13. $75m^{11}n^3$
14. $45p^{13}q^4$ 15. $120r^{10}s^{10}$ 16. $21t^{12}u^{10}$ 17. $96w^5v^6$
18. $21x^{18}y^3$ 19. $18e^{17}z^5$ 20. $28a^7b^{10}$ 21. $10c^{16}d^8$
22. $72e^6f^{16}$ 23. $84g^7h^{10}$ 24. $54i^7j^7$ 25. $320k^{20}l^7$
26. $90m^{15}n^7$ 27. $80o^{11}p^6$ 28. $100q^{21}r^9$ 29. $16s^9t^{15}$
30. $40u^{13}v^7$ 31. $144w^2x^{23}$ 32. $84x^{16}y^{11}$ 33. $144a^{16}b^4$

8-6 ▼ Working with Scientific Notation

1. 1.4×10^5 2. 3.56×10^{-7} 3. 4.23×10^8
4. 2.9×10^1 5. 9.36×10^{-2} 6. 8.15×10^1
7. 1.875×10^6 8. 3.8×10^{11} 9. 2.25×10^{-8}
10. 4.5×10^{-1} 11. 3.1×10^{10} 12. 5.9×10^{12}
13. 1.375×10^{23} 14. 4.6×10^{-12} 15. 8.01×10^{-22}
16. 1.65×10^2 17. 3.28×10^{11} 18. 1.3125×10^{18}
19. 1.225×10^{-8} 20. 5×10^{52} 21. 5.11875×10^{17}
22. 3.9×10^{12} 23. 1.95×10^7 bytes 24. 5.0625×10^{-3} g

8-7 ▼ Raising a Power to a Power

1. 4^{12} 2. 3^{18} 3. 5^8 4. 24^8 5. 9^{15} 6. 15^{20} 7. a^{14}
8. b^{29} 9. c^{17} 10. d^{43} 11. c^{34} 12. f^{42} 13. d^{16} 14. c^{37}
15. g^{68} 16. e^{19} 17. d^{38} 18. h^{100} 19. g^{28} 20. e^{36} 21. i^{102}
22. h^{40} 23. f^{38} 24. j^{82} 25. i^{39} 26. g^{37} 27. k^{56} 28. j^{19}
29. h^{54} 30. l^8 31. k^{86} 32. i^{60} 33. m^{15} 34. l^9 35. j^{14}
36. n^{64} 37. m^{49} 38. k^{18} 39. o^{75} 40. n^{27} 41. l^{22} 42. p^{42}
43. o^7 44. m^{58} 45. q^{98} 46. p^{28} 47. n^{16} 48. r^{58} 49. q^{53}
50. o^{52} 51. s^{30} 52. r^{93} 53. p^{37} 54. t^{18} 55. s^{39} 56. q^{126}
57. u^{39}

8-7 ▼ Raising a Product to a Power

1. a^{12} 2. b^{20} 3. c^{15} 4. d^{40} 5. c^{27} 6. g^{63} 7. $162a^{14}$
8. $40o^{20}$ 9. $128c^{12}$ 10. $1372b^{33}$ 11. $192p^{30}$ 12. $2430d^{20}$
13. $567c^{48}$ 14. $1280q^8$ 15. $256e^{24}$ 16. $352d^{15}$ 17. $1512r^9$
18. $375f^{24}$ 19. $2560e^{20}$ 20. $2s^{25}$ 21. $144g^{20}$ 22. $125x^{21}$
23. $500t^{12}$ 24. $1000h^{36}$ 25. $1152g^{18}$ 26. $1014u^6$
27. $16i^{56}$ 28. $3584h^{33}$ 29. $256v^8$ 30. $160j^{55}$ 31. $216i^{39}$
32. $10,000w^{21}$ 33. $324k^{20}$ 34. $405j^{60}$ 35. $729x^6$
36. $2560l^{27}$ 37. $363k^{24}$ 38. y^{25} 39. $144m^{14}$ 40. $108l^{24}$
41. $175z^{12}$ 42. $6000n^6$ 43. $1,000,000m^{36}$ 44. $750a^{27}$
45. $162o^{12}$ 46. $128n^8$ 47. $1125b^{20}$ 48. $8p^{15}$
49. 5.024×10^7 m^2

8-8 ▼ Dividing Powers with the Same Base

1. $\frac{1}{a^3}$ 2. b^2 3. $\frac{1}{c^2}$ 4. $\frac{1}{d^{20}}$ 5. $\frac{1}{c^4}$ 6. g^7 7. $\frac{b^8}{a}$

8. cd 9. e^2f^{10} 10. $\frac{1}{g^9}$ 11. h^2 12. $\frac{1}{i^7j^6}$ 13. $\frac{k^{10}}{l^4}$

14. $m^{17}n^{41}$ 15. $\frac{o^{51}}{p^{100}}$ 16. 2.5 gal 17. 5 h

8-8 ▼ Raising a Quotient to a Power

1. $-\frac{8}{27}$ 2. $-8f^6$ 3. k^2 4. $\frac{8}{27}$ 5. $\frac{27}{8y^2}$ 6. $\frac{x^2}{16}$

7. $\frac{49n^4}{64}$ 8. $\frac{9}{49}$ 9. $\frac{8p^6}{27}$ 10. $\frac{1}{27g^9}$ 11. $125r^3$

12. $\frac{y^6}{144}$ 13. $\frac{e^{12}}{4}$ 14. $-\frac{64}{125}$ 15. $\frac{121}{144t^2}$

9-1 ▼ Solving Equations Using the Pythagorean Theorem

1. ±3 2. ±3.87 3. ±3.16 4. ±3.46 5. ±2.83 6. ±4
7. 11.2 8. 20 9. 3.61 10. 7.81 11. 6.4 12. 5.83
13. 9.17 14. 5 15. 17 16. 8 17. 3, 4, 5 18. 13 ft
19. 127.28 ft

© Prentice-Hall, Inc.

9-1 ▼ Using the Converse

1. 5 **2.** 5.83 **3.** 10 **4.** 2.24 **5.** 15 **6.** 7.81
7. no; $5^2 + 7^2 \neq 10^2$ **8.** no; $7^2 + 9^2 \neq 11^2$
9. no; $3^2 + 8^2 \neq 12^2$ **10.** no; $4^2 + 7^2 \neq 12^2$
11. yes; $12^2 + 16^2 = 20^2$ **12.** yes; $6^2 + 8^2 = 10^2$
13. yes; $3^2 + 4^2 = 5^2$ **14.** no; $13^2 + 14^2 \neq 15^2$
15. yes; $15^2 + 20^2 = 25^2$ **16.** no; $6^2 + 8^2 \neq 14^2$
17. yes; $9^2 + 12^2 = 15^2$ **18.** no; $4^2 + 6^2 \neq 8^2$
19. no; $1^2 + 2^2 \neq 3^2$ **20.** no; $7^2 + 8^2 \neq 9^2$
21. yes; $18^2 + 24^2 = 30^2$ **22.** yes; $5^2 + 12^2 = 13^2$
23. no; $2^2 + 6^2 \neq 10^2$ **24.** no; $3^2 + 7^2 \neq 11^2$
25. no; $1.5^2 + 3^2 \neq 5^2$ **26.** yes; $4.5^2 + 6^2 = 7.5^2$
27. no; $5^2 + 7^2 \neq 9.5^2$ **28.** yes; $2.5^2 + 6^2 = 6.5^2$
29. no; $3.5^2 + 4.5^2 \neq 5.5^2$ **30.** no; $10.5^2 + 12^2 \neq 14^2$
31. no; $5^2 + 6.5^2 \neq 7^2$ **32.** no; $11^2 + 12^2 \neq 13.5^2$
33. no; $14^2 + 15^2 \neq 17.5^2$ **34.** no; $5^2 + 10^2 \neq 15^2$
35. no; $7.5^2 + 10^2 \neq 25^2$ **36.** no; $10^2 + 20^2 \neq 30^2$
37. yes; $10.5^2 + 14^2 = 17.5^2$ **38.** no; $5^2 + 15^2 \neq 25^2$
39. no; $24^2 + 25^2 \neq 26^2$ **40.** yes; $21^2 + 28^2 = 35^2$
41. no; $4.5^2 + 5^2 \neq 7^2$ **42.** no; $8^2 + 16^2 \neq 24^2$

9-2 ▼ Finding the Distance

1. 16 **2.** 25 **3.** 49 **4.** 81 **5.** 1 **6.** 4 **7.** 7.07 **8.** 4.47 **9.** 1
10. 6.4 **11.** 4.47 **12.** 5.1 **13.** 1.41 **14.** 7.07 **15.** 2
16. 3.16 **17.** 1.41 **18.** 2.83 **19.** 1.41 **20.** 2.83 **21.** 1.41
22. 3.61 **23.** 5.39 **24.** 1.41 **25.** 8.6 **26.** 1.41 **27.** 3.16
28. 2.83 **29.** 1.41 **30.** 2.83 **31.** 2 **32.** 5.1 **33.** 8.06
34. 50 yd **35.** $\approx$16.0 m

9-2 ▼ Using the Midpoint Formula

1. 1 **2.** 2 **3.** 4 **4.** 3 **5.** 1 **6.** 2 **7.** $(-1, 1)$ **8.** $(5, 4)$
9. $(4, 4)$ **10.** $(3, 4)$ **11.** $(0, 0)$ **12.** $(5, 5)$ **13.** $(0, 0)$
14. $(0, 3)$ **15.** $(4, 5)$ **16.** $(3, -3)$ **17.** $(0, 0)$ **18.** $(-4, 2)$
19. $(1, 3)$ **20.** $(8, 7)$ **21.** $(0, 3)$ **22.** $(0, 0)$ **23.** $(0, 4)$
24. $(0, 0)$ **25.** $(39.4N, 118.0W)$

9-3 ▼ Finding Trigonometric Ratios

1. 8 **2.** 15 **3.** 10 **4.** 8 **5.** 9 **6.** 8 **7.** $\frac{3}{5}$ **8.** $\frac{3}{5}$
9. $\frac{4}{3}$ **10.** $\frac{4}{5}$ **11.** $\frac{4}{3}$ **12.** $\frac{4}{5}$ **13.** $\frac{4}{5}$ **14.** $\frac{4}{5}$ **15.** $\frac{3}{5}$
16. $\frac{3}{5}$ **17.** $\frac{3}{5}$ **18.** $\frac{4}{3}$ **19.** $\frac{4}{5}$ **20.** $\frac{4}{5}$ **21.** $\frac{3}{4}$ **22.** $\frac{3}{4}$
23. $\frac{3}{4}$ **24.** $\frac{3}{5}$

9-3 ▼ Solving Problems Using Trigonometric Ratios

1. 0.1 **2.** 6.8 **3.** ~48.78 **4.** 8.32 **5.** 40 **6.** 9 **7.** 5.8
8. 33.7 **9.** 3.4 **10.** 47.0 m **11.** 64.4 **12.** 38.2 ft
13. 447.2 m

9-4 ▼ Multiplication with Radicals

1. 5 **2.** 13 **3.** 5 **4.** $5\sqrt{2}$ **5.** $\sqrt{41}$ **6.** $5\sqrt{2}$

7. $4\sqrt{2}$ **8.** $2\sqrt{5}$ **9.** $3\sqrt{3}$ **10.** $10\sqrt{3}$ **11.** $3\sqrt{2}$

12. $3\sqrt{5}$ **13.** $2\sqrt{2}$ **14.** $2\sqrt{3}$ **15.** $2\sqrt{7}$ **16.** $2\sqrt{6}$

17. $4\sqrt{3}$ **18.** $7\sqrt{2}$ **19.** $3\sqrt{6}$ **20.** $5\sqrt{6}$ **21.** $10\sqrt{10}$

22. $10\sqrt{2}$ **23.** $5\sqrt{3}$ **24.** $3\sqrt{7}$ **25.** $-2\sqrt{3}$ **26.** $5\sqrt{10}$

27. $6\sqrt{2}$ **28.** 4 **29.** 6 **30.** $2\sqrt{3}$ **31.** 30 **32.** $6\sqrt{10}$

33. $-30\sqrt{2}$ **34.** $45\sqrt{5}$ **35.** $5\sqrt{2}$ **36.** 18 **37.** 6

38. $2\sqrt{2}$ **39.** $10\sqrt{3}$ **40.** 6 **41.** $5\sqrt{10}$ **42.** $8\sqrt{3}$

43. $5\sqrt{2}$ **44.** $3\sqrt{2}$ **45.** $4\sqrt{13}$ **46.** $2\sqrt{6}$ m^2 **47.** 160 ft^2

9-4 ▼ Division with Radicals

1. 10 **2.** -9 **3.** 24 **4.** 22 **5.** 14 **6.** 14

7. $\frac{\sqrt{3}}{2}$ **8.** $\frac{\sqrt{7}}{3}$ **9.** $\frac{\sqrt{11}}{4}$ **10.** $\frac{7}{11}$ **11.** $\frac{2\sqrt{3}}{9}$ **12.** $\sqrt{7}$

13. $\frac{\sqrt{5}}{3}$ **14.** 2 **15.** $\frac{\sqrt{2}}{3}$ **16.** $\frac{\sqrt{3}}{4}$ **17.** $2\sqrt{3}$ **18.** 3

19. $\sqrt{13}$ **20.** $\frac{2\sqrt{5}}{9}$ **21.** $\frac{\sqrt{10}}{5}$ **22.** 3 **23.** 3 **24.** 5

25. 2 **26.** 1 **27.** $\frac{\sqrt{3}}{7}$ **28.** 2 **29.** $\frac{2}{5}$ **30.** $\frac{\sqrt{7}}{6}$ **31.** $\frac{\sqrt{10}}{9}$

32. $\frac{\sqrt{6}}{4}$ **33.** $\frac{\sqrt{7}}{5}$ **34.** $\frac{\sqrt{6}}{10}$ **35.** $\frac{\sqrt{11}}{12}$ **36.** $\frac{\sqrt{7}}{3}$

9-5 ▼ Simplifying Sums and Differences

1. $7x - 3$ **2.** $-11x - 5$ **3.** $-3y + 7$ **4.** $4y + 2$

5. $2y - 2$ **6.** $8y - 1$ **7.** $3\sqrt{3}$ **8.** $-3\sqrt{2}$ **9.** $2\sqrt{2}$

10. $7\sqrt{11}$ **11.** $-3\sqrt{2}$ **12.** $8\sqrt{7}$ **13.** $\sqrt{3}$ **14.** $4\sqrt{3}$

15. 0 **16.** $7\sqrt{3}$ **17.** $2\sqrt{7}$ **18.** $\sqrt{5}$ **19.** $-\sqrt{7}$ **20.** 0

21. -1 **22.** $2\sqrt{5}$ **23.** $6\sqrt{2}$ **24.** $4\sqrt{2}$ **25.** 0

26. $10\sqrt{5}$ **27.** $\sqrt{10}$ **28.** $7\sqrt{6}$ **29.** $\sqrt{10}$ **30.** $5\sqrt{5}$

31. $6\sqrt{2}$ **32.** $-3\sqrt{3}$ **33.** $-3\sqrt{6}$ **34.** $-3\sqrt{2}$

35. $21\sqrt{3}$ **36.** $41\sqrt{2}$ **37.** $19\sqrt{2}$ **38.** $9\sqrt{2} + 9$

39. $-3\sqrt{3}$ **40.** $5\sqrt{3}$ **41.** $-5\sqrt{5}$ **42.** $5\sqrt{5}$ **43.** $\sqrt{2}$

44. $\sqrt{2}$ **45.** $-\sqrt{5}$

Two-Year Algebra Handbook Answers (continued)

9-5 ▼ Simplifying Products, Sums, and Differences

1. $4x - xy$ **2.** $24 + 6y$ **3.** $9y - xy$ **4.** $3x + 6$

5. $5x + 15$ **6.** $5x + xy$ **7.** $3\sqrt{6} + 3$ **8.** $15 - 6\sqrt{2}$

9. $4 - 6\sqrt{2}$ **10.** $5\sqrt{2} + 6\sqrt{3}$ **11.** $10\sqrt{3} + 6\sqrt{6}$

12. $-12 + 6\sqrt{3}$ **13.** 30 **14.** $3\sqrt{5} - 10$

15. $-5\sqrt{3} - 6$ **16.** $4\sqrt{3} - 3\sqrt{2}$ **17.** $4\sqrt{3} + 3$

18. $6\sqrt{2} - 8\sqrt{3}$ **19.** $5\sqrt{7} + 3\sqrt{35}$ **20.** $5 - 3\sqrt{5}$

21. $6 - 2\sqrt{6}$ **22.** $2\sqrt{2} + 2$ **23.** $2\sqrt{6} - 6\sqrt{2}$

24. $2\sqrt{2} - 6\sqrt{3}$ **25.** 30 **26.** $2\sqrt{6} - \sqrt{15}$

27. $-3\sqrt{6} + 9$ **28.** $\sqrt{6} + \sqrt{10}$ **29.** 15

30. $8\sqrt{5} + 60$ **31.** $15\sqrt{15} - 15$ **32.** $4\sqrt{6} + 12$

33. $3\sqrt{5} + 5\sqrt{2}$ **34.** $\sqrt{6} + 3\sqrt{2}$

35. $\sqrt{14} + 5\sqrt{7}$ **36.** $4\sqrt{2} + 4$ **37.** $5\sqrt{3} + 3\sqrt{2}$

38. $3\sqrt{6} + 2\sqrt{3}$ **39.** $3\sqrt{2} + \sqrt{15}$ **40.** $5\sqrt{2} - 2$

41. $7\sqrt{7} - 7$ **42.** $\sqrt{6} - 3\sqrt{2}$ **43.** $4\sqrt{3} - 12$

44. $-2\sqrt{3}$ **45.** $10\sqrt{3} - 40$ **46.** $\sqrt{3} + 2$ **47.** $\sqrt{6} + 3$

48. $5\sqrt{7}$ **49.** 0 **50.** 0 **51.** $4\sqrt{5} - \sqrt{10}$

9-6 ▼ Solving a Radical Equation

1. 13 **2.** 3 **3.** 6 **4.** 6 **5.** 7 **6.** $2\sqrt{2}$ **7.** 18 **8.** 20 **9.** 11
10. 32 **11.** 4 **12.** 4 **13.** 81 **14.** 8 **15.** 5 **16.** 16 **17.** 6
18. 49 **19.** -1 **20.** 15 **21.** -1 **22.** 9 **23.** 1296 **24.** 14
25. 5 **26.** 1 **27.** 25 **28.** 81 **29.** 5 **30.** 4 **31.** no solution
32. 36 **33.** 1

9-6 ▼ Solving Equations with Extraneous Solutions

1. $6\sqrt{2}$ **2.** $-3\sqrt{3}$ **3.** $-5\sqrt{x}$ **4.** $5\sqrt{3}$ **5.** $3\sqrt{3}$

6. $-3\sqrt{b}$ **7.** 3 **8.** no solution **9.** $\sqrt{5}$ **10.** no solution
11. no solution **12.** 11 **13.** 5 **14.** 2 **15.** no solution
16. no solution **17.** 3 **18.** 9 **19.** 2 **20.** 4 **21.** $2, 4$ **22.** 1
23. 4 **24.** 2

9-7 Graphing Square Root Functions

1. 0 **2.** 15 **3.** 2 **4.** 6 **5.** 4 **6.** 5

7. $x \geq 0$; 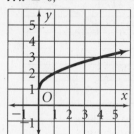 **8.** $x \geq -2$;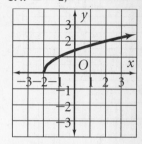

9. $x \geq 6$; 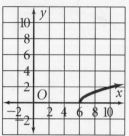 **10.** $x \geq -2$;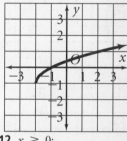

11. $x \geq -3$; 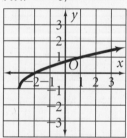 **12.** $x \geq 0$;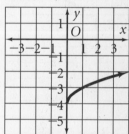

13. $x \geq 7$; 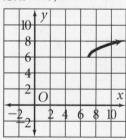 **14.** $x \geq 5$;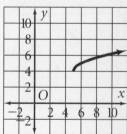

15. $x \geq -1$;

16. $x \geq -4$;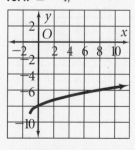

17. $x \geq 0$; **18.** $x \geq 1$;

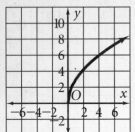

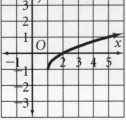

27. $x \geq -1$;

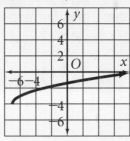

28. $x \geq -7$;

19. $x \geq 3$;

20. $x \geq -5$;

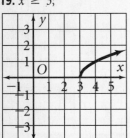

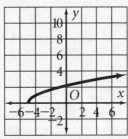

29. $x \geq 0$;

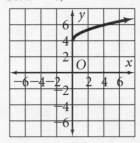

30. $x \geq 0$;

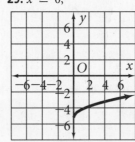

21. $x \geq 0$;

22. $x \geq -6$;

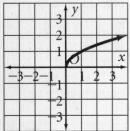

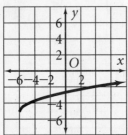

31. $x \geq 0$;

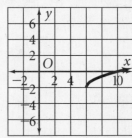

32. $x \geq 6$;

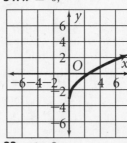

23. $x \geq 2$;

24. $x \geq 3$;

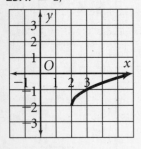

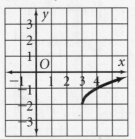

33. $x \geq 0$;

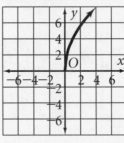

34. $x \geq 3$;

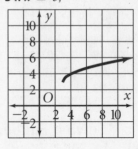

25. $x \geq 4$;

26. $x \geq 5$;

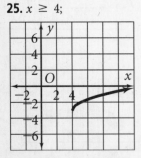

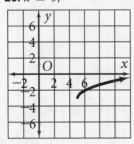

35. $x \geq 3$;

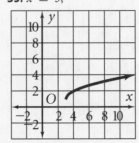

36. $x \geq 4$;

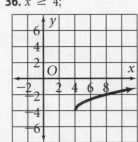

37. $x \geq 4$;

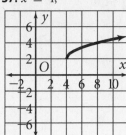

38. $x \geq 0$;

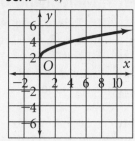

39. $x \geq 2$;

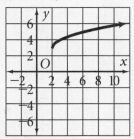

40. $x \geq 0$;

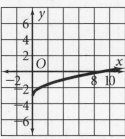

41. $x \geq -3$;

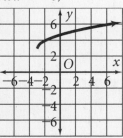

42. $x \geq 0$;

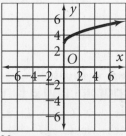

43. $x \geq 0$;

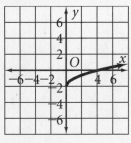

44. $x \geq 0$;

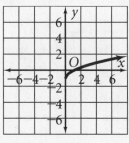

45. $x \geq 0$;

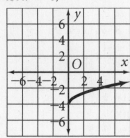

9-8 Analyzing Data
Using Standard Deviation

1. 2.33 **2.** 12 **3.** 5 **4.** 85 **5.** 1 **6.** 60 **7.** 1.4 **8.** 1.8
9. 4.5 **10.** 1.36 **11.** 3.87 **12.** 3.6 **13.** 7.5 **14.** 4 **15.** 0.7
16. 0.67 **17.** 14.1 **18.** 13.8 **19.** 14.1 **20.** 0.11 **21.** 2.9
22. 5.6 **23.** 1.1 **24.** 9

10-1 ▼ Describing Polynomials

1. 5, 2, −3, −4 **2.** 9, 3, 1, −7 **3.** 0, −1, −3, −5 **4.** 14, 12,
−12, −14 **5.** 8, 2, −3, −5 **6.** 10, 7, −8, −9 **7.** $4x^2 + 7x$;
quadratic binomial **8.** $9x - 5$; linear binomial
9. $5x^2 - 12x + 10$; quadratic trinomial
10. $-3y^3 + 8y^2 + y - 1$; cubic polynomial with four terms
11. $k^6 + 3k^4 - 2$; sixth degree trinomial
12. $7a^7 - 3a^3 - 2a - 12$; seventh degree polynomial
with four terms **13.** $-b^3 + b^2 + 5b$; cubic trinomial
14. $11c^3 + 2$; cubic binomial **15.** $3g^2 - 5g - 4$;
quadratic trinomial **16.** $13c^5 + 12c^2$ $17c$; fifth degree
trinomial **17.** $6x^2 - x - 3$; quadratic trinomial
18. $5y^5 + 2y^2 - y - 9$; fifth degree polynomial with four
terms **19.** $-12x^3 + 5x + 14$; cubic trinomial
20. x^3; cubic monomial **21.** $y + 4$; linear binomial
22. $y5 - 5$; fifth degree binomial **23.** $4x^3 - x^2 - 6x + 5$;
cubic polynomial with four terms **24.** $4p^2 - 5p + 8$;
quadratic trinomial **25.** $2p^2 - 6$; quadratic binomial
26. y^4; fourth degree monomial **27.** $x^{14} + x^{10} + x^4$; 14th
degree trinomial **28.** $b^3 + 95$; cubic binomial
29. $x^4 + x^2 - 3$; fourth degree trinomial
30. $g^3 + 4g^2 + g - 3$; cubic polynomial with four terms
31. $-4x^5 + 25$; fifth degree binomial
32. x; linear monomial **33.** $-8a^{10} + a^7 - a^5 + 5a^2 - 7$;
tenth degree polynomial with five terms

10-1 ▼ Adding Polynomials

1. $3x - 2y$ **2.** $c - d$ **3.** $3x + 2$ **4.** $-7c + 7d + 2$
5. $15a - 1$ **6.** $-7t + 2u - 5$ **7.** $5x^2 + 3x$
8. $x^2 + 4x + 3$ **9.** $2y^2 - 4y - 1$
10. $4a^3 + 4a^2 - 9a - 1$ **11.** $4b^2 + 1$
12. $5x^2 + 2x + 3$ **13.** $-2x$ **14.** $2y^2 - 4y - 4$

10-1 ▼ Subtracting Polynomials

1. -4 **2.** $12 - a$ **3.** $-x + y + z$ **4.** $x - 3$
5. $-4 + b + c$ **6.** $5 + 2m$ **7.** $2x^3 + 9x + 1$
8. $6x^4 - 3x^2 + 4x - 3$ **9.** $5y^2 - 10y + 1$
10. $2a^3 - a^2 + 5a + 5$ **11.** $-4b^2 - 3$
12. $3x^3 - x^2 - 3x + 12$ **13.** $2y^3 - y + 14$
14. $-x^2 + 6x + 10$

10-2 ▽ Multiplying by a Monomial

1. $3x + 6$ **2.** $5a - 10$ **3.** $2x + 2y$ **4.** $2m - 14$
5. $-b + c$ **6.** $5x - 20$ **7.** $10x^2 - 6x$
8. $2x^4 + 10x^3 - 6x^2$ **9.** $-4y^3 - y^2$
10. $3a^4 + 12a^3 + 21a^2$ **11.** $-4b^6 + 20b^5 - 8b^2$
12. $-3x^4 - 3x^3 + 12x$ **13.** $3y^3 + 6y^2 - 21y$
14. $2x^4 - 4x^3 - 5x^2$ **15.** $4x^2 - 16x$ **16.** $-2x^2 + x$
17. $-y^3 + 3y^2$ **18.** $4b^2 - 4b$ **19.** $-c^4 - 4c^3$
20. $x^5 - x^4$ **21.** $a^4 + a^2$ **22.** $n^3 - 4n$ **23.** $y^2 - y$
24. $-5x^2 - 10x$

10-2 ▿ Factoring Out a Monomial

1. 5 **2.** 7 **3.** 2 **4.** 4 **5.** 3 **6.** 6 **7.** $5x(2x - 5)$
8. $3(3x + 4)$ **9.** $3y(2y^2 - 5y + 1)$
10. $7a(2a^2 + 5a + 3)$ **11.** $b^2(-b^2 + 5b - 4)$
12. $17x(3x^2 + x - 2)$ **13.** $y(3y^2 + 2y - 7)$
14. $4x^2(3x^3 + x^2 - 9x + 6)$ **15.** $6(b + 3)$
16. $7(y + 3)$ **17.** $x(4x + 1)$ **18.** $8y^2(y + 3)$
19. $7a(a + 7)$ **20.** $5w^3(w - 4)$ **21.** $2p(2p^2 - 6p + 9)$
22. $15t^3(t^5 - 3)$ **23.** $14a^2(a^2 - 3a + 2)$
24. $6x(2x^2 - 3x + 4)$ **25.** $4(4x^2 - 3x + 6)$
26. $y^2(3y^2 - 4y + 2)$ **27.** $4(x^2 + 3x + 2)$
28. $a(a + 7)$ **29.** $2p(p^2 + 2p + 1)$ **30.** $8(x - y)$
31. $3(k^2 - 3)$ **32.** $2n(4n - 5)$ **33.** $4(x^2 + 2x + 3)$
34. $3x(x^2 + 2x - 3)$ **35.** $3(p + 2)$ **36.** $5(2z + 3)$
37. $4(2k - 1)$ **38.** $9(3r + 1)$ **39.** $x(7x + 2)$
40. $y(14 - y^3)$ **41.** $12(1 - a)$ **42.** $2x^2(x - 6)$
43. $5a^2(1 - 3a)$ **44.** $7c(1 + 2c)$ **45.** $6x^2(x^2 + 3)$
46. $13x^2(x + 2)$ **47.** $11w^2(3w + 1)$ **48.** $a^2(7 - 9a^2)$
49. $2j(2j^2 - 3j + 4)$ **50.** $5(n - 3n^2 + 2)$
51. $4a(a + 2a^2 + 3)$ **52.** $x(7x^2 + 6x + 5)$
53. $3(4m^2 + 9m^3 - m - 2)$
54. $2r(4r^4 - 2r^3 + 16r + 3)$
55. $-3a(a - 5a^2 + 6)$ **56.** $-7(3x^4 - 5x^2 - 2x + 4)$
57. $3a(a - 1)$

10-3 ▽ Multiplying Two Binomials

1. $2x^2 - 6x$ **2.** $11y$ **3.** $-3a^2 - 6a$ **4.** $3b^2 - b$ **5.** $3xy$
6. $-7x^2 + 9x$ **7.** $3a^2 - 13a - 10$ **8.** $x^2 - 11x + 28$
9. $6y^2 + 11y + 5$ **10.** $2a^2 + a - 15$
11. $2c^2 + 3bc - 2b^2$ **12.** $16x^2 + 2x - 5$
13. $-20y^2 + 16y + 21$ **14.** $6x^2 + x - 35$
15. $3p^2 - 5p + 2$ **16.** $x^2 + x - 2$ **17.** $y^2 + 5y + 6$
18. $14 + 5x - x^2$ **19.** $-y^2 + 18y - 81$
20. $k^2 + 7k + 10$ **21.** $x^2 + x - 12$
22. $6y^2 - 13y + 5$ **23.** $110 + 21x - x^2$
24. $4y^2 - 5y - 21$

10-3 ☑ Multiplying Using FOIL

1. $x^2 - 4x$ **2.** $5y$ **3.** $-3a$ **4.** $5b$ **5.** y **6.** $-4x^2 + 3x$
7. $y^2 - 3y - 4$ **8.** $x^2 + 8x + 15$ **9.** $x^2 - 3y + 2$
10. $2a^2 - 5a + 2$ **11.** $c^2 + 7c + 12$
12. $2x^2 - 12x + 10$ **13.** $-y^2 + 2y + 3$
14. $2x^2 + 12x + 10$ **15.** $6x^2 - 23x - 4$ **16.** $4x^2 - 16$
17. $x^2 - 3x + 2$ **18.** $x^2 + 7x + 12$ **19.** $x^2 - x - 30$
20. $4x^2 + 3x - 1$ **21.** $5x^2 + 8x - 4$
22. $3y^2 - 4y + 1$ **23.** $y^2 + y - 90$
24. $4x^2 + 3x - 10$ **25.** $2 + x - x^2$
26. $20 - y - y^2$ **27.** $2y^2 + 7y + 5$
28. $3x^2 + 8x - 3$ **29.** $9x^2 - 8x - 1$
30. $40 + 14x + x^2$ **31.** $24x^2 - 6x - 3$
32. $x^2 - 7x + 12$ **33.** $2y^2 + 7y - 4$
34. $5y^2 - 13y - 6$ **35.** $6x^2 - 5x - 6$
36. $3x^2 + 4x - 4$ **37.** $4x^2 + 15x - 25$
38. $14 - 19x - 3x^2$ **39.** $15 + 38x + 24x^2$
40. $3 + 10x + 8x^2$ **41.** $x^2 - 17x + 30$
42. $y^2 + 7y - 30$ **43.** $y^2 + y - 42$
44. $x^2 - 15x + 56$ **45.** $6y^2 - 11y - 2$
46. $3x^2 + 7x - 20$ **47.** $2b^2 + 3b - 27$
48. $63 - 2x - x^2$

10-3 ☒ Multiplying a Trinomial and a Binomial

1. $3x - 9$ **2.** $4y + 28$ **3.** $-3a - 6$ **4.** $20 - 4b$
5. $4x - 4y$ **6.** $-27 + 9a$ **7.** $2a^3 - 3a^2 - 17a + 30$
8. $x^3 - 2x^2 - 13x + 20$ **9.** $-y^3 + 4y^2 + 12y + 7$
10. $3a^3 - 4a^2 - 49a - 30$ **11.** $2c^3 + c^2 - 2c + 8$
12. $2x^3 + 3x^2 - 12x + 5$ **13.** $y^3 - 6y^2 - 29y - 6$
14. $-4x^3 - 9x^2 + 3x + 10$ **15.** $x^3 - 5x^2 + 7x - 3$
16. $3x^3 - x^2 - 5x - 2$ **17.** $2y^3 - 17y^2 - 11y + 18$
18. $y^3 + y^2 - 7y + 20$ **19.** $x^3 + 4x^2 - 17x + 28$
20. $3y^3 + 5y^2 + 3y + 1$ **21.** $9y^3 - 39y^2 + 12y - 8$
22. $10y^3 - 55y^2 + 26y - 5$ **23.** $x^3 - 2x^2 + 1$
24. $x^2 + 6x^2 + 11x + 30$

10-4 ▽ Using Tiles

1. -15 **2.** 16 **3.** 18 **4.** 0 **5.** 8 **6.** 8 **7.** $(x + 9)(x - 1)$
8. $(x - 3)(x + 2)$ **9.** $(x + 5)(x + 1)$
10. $(x - 2)(x - 2)$ **11.** $(x - 3)(x - 1)$
12. $(x + 7)(x - 1)$ **13.** $(x - 2)(x + 1)$
14. $(x + 2)(x + 4)$ **15.** $(x + 3)(x + 3)$
16. $(x - 3)(x - 4)$ **17.** $(x - 2)(x - 3)$
18. $(x - 4)(x + 5)$ **19.** $(x - 3)(x - 5)$
20. $(x - 2)(x - 8)$ **21.** $(x + 1)(x + 3)$
22. $(x - 1)(x - 1)$ **23.** $(x + 2)(x + 1)$
24. $(x + 2)(x - 1)$ **25.** $(x + 2)(x + 2)$
26. $(x + 4)(x + 1)$ **27.** $(x + 1)(x + 1)$
28. $(x - 2)(x - 4)$ **29.** $(x + 5)(x - 3)$
30. $(x - 4)(x + 1)$ **31.** $(x - 1)(x + 8)$

32. $(x + 5)(x + 2)$ **33.** $(x - 1)(x - 10)$
34. $(x - 2)(x + 4)$ **35.** $(x - 6)(x + 2)$
36. $(x - 7)(x + 1)$ **37.** $(x + 7)(x + 1)$
38. $(x - 10)(x + 1)$ **39.** $(x + 4)(x + 5)$

10-4 ▽ Testing Possible Factors

1. 3 **2.** 4 **3.** 3 **4.** 7 **5.** 2 **6.** 5 **7.** 2 **8.** 3 **9.** 3 **10.** 2
11. 4 **12.** 3 **13.** 1 **14.** 6 **15.** 2 **16.** 6 **17.** 8 **18.** 5
19. 5 **20.** 4 **21.** 8 **22.** 2

10-4 ▽ Factoring Trinomials

1. 1, 2, 4, 8, 16 **2.** 1, 3, 7, 21 **3.** 1, 2, 3, 5, 6, 10, 15, 30
4. 1, 2, 7, 14 **5.** 1, 2, 3, 4, 6, 8, 12, 24 **6.** 1, 3, 5, 15
7. $(x + 2)(2x - 9)$ **8.** $2(x - 2)(x + 3)$
9. $(3x + 5)(x - 4)$ **10.** $(5x - 7)(x + 2)$
11. $(3x + 5)(x - 3)$ **12.** $(2x - 3)(3x + 7)$
13. $2(x - 2)(x + 4)$ **14.** $(7x - 9)(x + 1)$
15. $(2x + 7)(x + 2)$ **16.** $(2x - 5)(2x - 1)$
17. $(3x - 4)(2x + 5)$ **18.** $(8x + 3)(x - 2)$
19. $(2x + 1)(x + 1)$ **20.** $(5x - 2)(x - 1)$
21. $(2x - 1)(3x + 5)$ **22.** $(4x - 1)(x - 2)$
23. $(4x - 3)(3x + 8)$ **24.** $(5x + 12)(2x + 3)$
25. $(3x - 1)(4x - 1)$ **26.** $(9x + 4)(2x - 7)$
27. $(9x - 2)(x + 3)$ **28.** $(2x - 3)(x + 2)$
29. $(3x + 1)(x + 2)$ **30.** $(2x + 7)(x - 11)$
31. $(5x - 1)(3x + 1)$ **32.** $(3x - 7)(4x + 5)$
33. $(7x - 3)(2x - 1)$ **34.** $(7x - 3)(2x + 3)$
35. $(3x - 13)(x + 2)$ **36.** $(2x + 1)(2x + 1)$
37. $(5x + 7)(3x + 1)$ **38.** $(5x - 3)(x + 2)$
39. $(5x + 1)(x + 1)$ **40.** $(3x - 1)(x - 5)$
41. $(5x - 2)(x - 4)$ **42.** $(3x - 5)(3x - 5)$
43. $(4x + 3)(x - 3)$ **44.** $(3x - 1)(x - 4)$
45. $(3x + 1)(2x + 1)$ **46.** $(7x - 2)(x + 7)$
47. $(2x - 3)(3x + 1)$ **48.** $(5x - 4)(x + 1)$

10-5 ▽ Factoring a Difference of Two Squares

1. -2 **2.** 0 **3.** 8 **4.** 11 **5.** 0 **6.** 7 **7.** $(a - 9)(a + 9)$
8. $(x - 4)(x + 4)$ **9.** $(a - 5)(a + 5)$
10. $(x - 8)(x + 8)$ **11.** $(x - 1)(x + 1)$
12. $(b - 6)(b + 6)$ **13.** $(2c - 11)(2c + 11)$
14. $(3c + 10)(3c - 10)$ **15.** $(x + 3)(x - 3)$
16. $(y - 13)(y + 13)$ **17.** $(b + 2)(b - 2)$
18. $(x - 15)(x + 15)$ **19.** $(c - 12)(c + 12)$
20. $(2c - 7)(2c + 7)$ **21.** $(4y - 8)(4y + 8)$
22. $(3c - 9)(3c + 9)$ **23.** $(x - y)(x + y)$
24. $(2x - 2)(2x + 2)$ **25.** $(4x + 3)(4x - 3)$
26. $(10x + 5)(10x - 5)$ **27.** $(6y + 4)(6y - 4)$
28. $(7y + 8)(7y - 8)$ **29.** $(11b - 2c)(11b + 2c)$
30. $(2x + 7)(2x - 7)$ **31.** $(10x - 12)(10x + 12)$
32. $(8x - 5)(8x + 5)$ **33.** $(5b - 1)(5b + 1)$

34. $(4x + 9)(4x - 9)$ **35.** $(5x - 10)(5x + 10)$
36. $(4x + 5)(4x - 5)$ **37.** $(12x - 6)(12x + 6)$
38. $(b - c)(b + c)$ **39.** $(7x + 2)(7x - 2)$
40. $(8b - 7)(8b + 7)$ **41.** $(6x + 1)(6x - 1)$
42. $(4x - 6)(4x + 6)$ **43.** $(2c - 12)(2c + 12)$
44. $(9x + 10)(9x - 10)$ **45.** $(3c - 4)(3c + 4)$
46. $(5b - 13)(5b + 13)$ **47.** $(12x - 11)(12x + 11)$
48. $(6x - 2)(6x + 2)$ **49.** $(3x + 5)(3x - 5)$
50. $(3b - 9)(3b + 9)$ **51.** $(8b + 1)(8b - 1)$
52. $(5y + 3)(5y - 3)$ **53.** $(2b - 1)(2b + 1)$
54. $(10x + 20)(10x - 20)$

10-5 ▽ Factoring a Perfect Square Trinomial

1. 10 **2.** -4 **3.** -6 **4.** 1 **5.** 3 **6.** -2 **7.** $(x - 7)^2$
8. $(y + 5)^2$ **9.** $(x - 11)^2$ **10.** $(y + 3)^2$ **11.** $(x - 2)^2$
12. $(y + 4)^2$ **13.** $(y - 10)^2$ **14.** $(x - 12)^2$ **15.** $(x - 8)^2$
16. $(y + 1)^2$ **17.** $(x - 6)^2$ **18.** $(2y - 2)^2$ **19.** $(5y - 1)^2$
20. $(4x - 3)^2$ **21.** $(3x - 5)^2$ **22.** $(y - 15)^2$
23. $(y - 1)^2$ **24.** $(2x - 3)^2$ **25.** $(x + 2)^2$
26. $(3y - 6)^2$ **27.** $(x + 7)^2$ **28.** $(y + 20)^2$
29. $(5x - 6)^2$ **30.** $(2x + 3)^2$ **31.** $(4y - 1)^2$
32. $(3x + 2)^2$ **33.** $(5y - 3)^2$ **34.** $(y + 10)^2$
35. $(2x - 9)^2$ **36.** $(3x - 1)^2$ **37.** $(x + 11)^2$
38. $(2y + 2)^2$ **39.** $(y + 12)^2$ **40.** $(x - 4)^2$
41. $(y - 5)^2$ **42.** $(y - 3)^2$

10-6 Solving Equations by Factoring

1. 6 **2.** 28 **3.** 40 **4.** 26 **5.** 49 **6.** 18 **7.** $-3, 8$ **8.** 3, 6
9. $-4, -3$ **10.** $-5, 6$ **11.** 6 **12.** $-6, -2$ **13.** $3.5, -3.5$
14. $-17, 1$ **15.** $-2, 3$ **16.** 0, 7 **17.** $2, -6$ **18.** $-6, -1$
19. $6, -2$ **20.** -4 **21.** $-4, 2$ **22.** $10, -9$ **23.** 5, 3
24. $0, -5$ **25.** 0, 2 **26.** $0, -4$ **27.** $7, -3$ **28.** $-4, 2$
29. $6, -3$ **30.** $-1, 10$ **31.** 0, 4 **32.** $0, -7$ **33.** $-3, 4$
34. $-5, 2$ **35.** $4, -8$ **36.** $-2, 9$ **37.** $-21, 1$ **38.** 5, 8
39. 5, 6 **40.** $6, -1$ **41.** $-3, 8$ **42.** 0, 5 **43.** $-1, -24$
44. 3, 9 **45.** 2, 8 **46.** 1, 8 **47.** 3, 12 **48.** $-1, -16$
49. ± 9 **50.** 9 **51.** $-2, -7$

10-7 Choosing an Appropriate Method for Solving

1. 5 **2.** -4 **3.** -2 **4.** -3 **5.** -2 **6.** 4 **7.** $1.5, -4$

8. $\frac{5}{3}, 1$ **9.** $-\frac{4}{3}$ **10.** $\frac{2}{5}, -1$ **11.** ± 4 **12.** $-\frac{5}{2}$

13. $-1, -6$ **14.** $-\frac{1}{4}, 1$ **15.** $\frac{1}{3}, -1$ **16.** ± 2

17. $\frac{4}{9}, -1$ **18.** -1

11-1 ▼ Solving Inverse Variations

1. 8 **2.** 5 **3.** 9 **4.** 7 **5.** 5 **6.** 4 **7.** 15 **8.** 6 **9.** 3 **10.** 4
11. 1.6 h **12.** 42 kg **13.** 10

11-1 ▼ Comparing Direct and Inverse Variation

1. 6 **2.** 3 **3.** 6 **4.** 1 **5.** inverse; $xy = 24$
6. direct; $y = 2x$ **7.** direct; $y = 3x$ **8.** inverse; $xy = 36$
9. direct; $y = 6x$ **10.** direct; $y = 4x$

11-2 ▼ Exploring Rational Functions

1. 4 **2.** 8 **3.** 6 **4.** -2 **5.** 4 **6.** 16 **7.** $-\frac{2}{3}$ **8.** -6

9. $-\frac{6}{5}$ **10.** $-\frac{2}{3}$ **11.** $\frac{2}{9}$ **12.** -5 **13.** $-\frac{1}{3}$ **14.** -1

15. $\frac{1}{3}$ **16.** 1 **17.** $-\frac{1}{4}$ **18.** 1 **19.** $-\frac{4}{7}$ **20.** $\frac{1}{2}$ **21.** 3

22. 7 **23.** -1 **24.** -1 **25.** $-\frac{15}{2}$ **26.** 4 **27.** -3 **28.** 3

29. 4 **30.** $\frac{3}{4}$ **31.** $\frac{2}{7}$ **32.** $\frac{25}{7}$ **33.** 5 **34.** -1 **35.** 4

36. $\frac{5}{2}$ **37.** -2 **38.** -1 **39.** $-\frac{11}{3}$

11-2 ▼ Graphing Rational Functions

1. -12 **2.** 0 **3.** 0, -5 **4.** ±3 **5.** 7 **6.** 1.8
7. 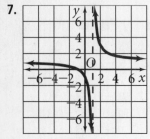 **8.**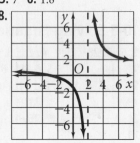

9. 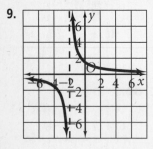 **10.**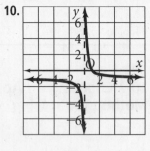

11. 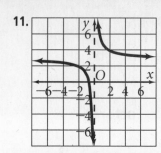 **12.**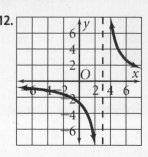

13. 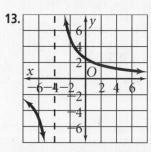 **14.**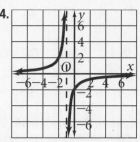

15. 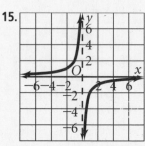 **16.**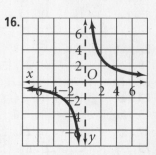

17. 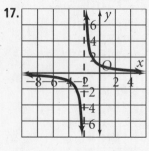 **18.**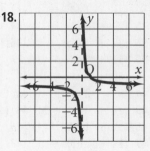

19. 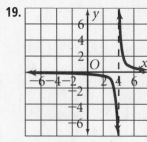 **20.**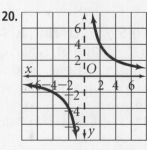

21.

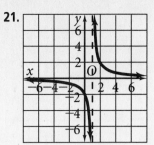

11-3 ▼ Simplifying Rational Expressions

1. $2x^2$ **2.** $2x$ **3.** 2 **4.** x **5.** $\frac{x}{7} + 1$ **6.** 1 **7.** $\frac{1}{b + 2}; -2$

8. $2; -3$ **9.** $\frac{2}{b}; 0$ **10.** $\frac{5}{3}; 4$ **11.** $\frac{1}{x - 1}; \pm 1$ **12.** $\frac{a}{2}; 0$

13. $\frac{x + 2}{x - 2}; 2$ **14.** $\frac{1}{2}; 1$ **15.** $\frac{1}{x - 4}; \pm 4$

16. $\frac{1}{x - 1}; 1, 0$ **17.** $-\frac{2}{x}; 0, 3$ **18.** $\frac{x + 4}{x + 3}; -3, 5$

19. $7; -3$ **20.** $\frac{2}{2x - 1}; \frac{1}{2}, 3$ **21.** $2x; 0$ **22.** $-\frac{x}{2}; 2$

23. $\frac{2}{x + 3}; \pm 3$ **24.** $2; 2$ **25.** $5; 3$ **26.** $-1; 1$

27. $2x + 1; 3$ **28.** $3; -4$ **29.** $-x + 2; 3$ **30.** $-x - 1; 1$

11-3 ▼ Multiplying and Dividing Rational Expressions

1. 2 **2.** -5 **3.** $-\frac{5}{4}$ **4.** $\frac{5}{3}$ **5.** $-\frac{1}{8}$ **6.** 4 **7.** $\frac{2}{x - 1}$

8. $\frac{7x}{2}$ **9.** $\frac{3}{2}$ **10.** $\frac{5}{3}$ **11.** $2(x + 2) = 2x + 4$

12. $18x$ **13.** 4 **14.** $3x$ **15.** $\frac{x^2}{12}$ **16.** 1 **17.** 2 **18.** x

19. 1 **20.** $\frac{1}{2}$ **21.** 6 **22.** $4x$ **23.** 3 **24.** $6x + 6$

11-4 Operations with Rational Expressions

1. 4 **2.** x **3.** $2 \cdot x$ **4.** $4x$ **5.** $2x$ **6.** $3 \cdot x$ **7.** $\frac{1}{4x}$

8. $\frac{4x + 15}{5x}$ **9.** $\frac{14 + 3x}{4x}$ **10.** $\frac{9 - 5x}{x^2}$ **11.** $\frac{21}{8}$

12. $\frac{2x + 3}{5}$ **13.** $\frac{6x - 1}{7}$ **14.** $\frac{2 - 7x}{14}$ **15.** $\frac{11}{4b}$

16. $\frac{1}{7k}$ **17.** $\frac{4}{5c}$ **18.** $\frac{3x + 4}{x^2}$ **19.** $\frac{35 - x^3}{5x^2}$

20. $\frac{x - 1}{x^2}$ **21.** $-\frac{11}{9}$ **22.** 1 **23.** $\frac{2x^2 + 3}{4x}$ **24.** $\frac{9}{10}$

25. $\frac{8}{x}$ **26.** $-\frac{3}{2x}$ **27.** $\frac{31}{10}$

11-5 Solving Rational Equations

1. $2\frac{1}{2}$ **2.** $8\frac{1}{8}$ **3.** $11\frac{1}{2}$ **4.** $\frac{1}{6}$ **5.** $1\frac{3}{4}$ **6.** $4\frac{10}{11}$

7. $5, -2$ **8.** $-\frac{1}{2}, \frac{4}{3}$ **9.** $1 \cdot 3$ **10.** $-1, \frac{2}{3}$ **11.** $-4, 2$

12. $\frac{4}{3}$ **13.** $-2, 5$ **14.** $-\frac{3}{4}$ **15.** 2 **16.** 3 **17.** -1

18. -3 **19.** 3 **20.** 12 **21.** 0

11-6 ▼ Using the Multiplication Counting Principle

1. 15 **2.** 32 **3.** 110 **4.** 42 **5.** 117 **6.** 24 **7.** 260 **8.** 6
9. 28 **10.** 168 **11.** 20 **12.** 56 **13.** 300 **14.** 121

11-6 ▼ Finding Permutations

1. 60 **2.** 210 **3.** 720 **4.** 1320 **5.** 24 **6.** 120 **7.** $362,880$
8. 120 **9.** 5040 **10.** 720 **11.** 720 **12.** $360,360$ **13.** 42
14. 90 **15.** 120 **16.** 24 **17.** 24 **18.** 60 **19.** 720 **20.** 210
21. 120

11-7 Combinations

1. 20 **2.** 210 **3.** 95040 **4.** 4 **5.** 5040 **6.** 30 **7.** 792 **8.** 35
9. 10 **10.** 35 **11.** 15